Water Resources Development and Management

More information about this series at http://www.springer.com/series/7009

Asit K. Biswas · Cecilia Tortajada
Editors

Water Security, Climate Change and Sustainable Development

Springer

Editors
Asit K. Biswas
Lee Kuan Yew School of Public Policy
National University of Singapore
Singapore
Republic of Singapore

Cecilia Tortajada
Institute of Water Policy, Lee Kuan Yew School of Public Policy
National University of Singapore
Singapore
Republic of Singapore

ISSN 1614-810X ISSN 2198-316X (electronic)
Water Resources Development and Management
ISBN 978-981-287-974-5 ISBN 978-981-287-976-9 (eBook)
DOI 10.1007/978-981-287-976-9

Library of Congress Control Number: 2015951397

Springer Singapore Heidelberg New York Dordrecht London

Printed on acid-free paper

Springer Science+Business Media Singapore Pte Ltd. is part of Springer Science+Business Media (www.springer.com)

Foreword

The Vibrant Gujarat series of biennial summits was the brainchild of my predecessor and the current Prime Minster of India, Hon. Narendra Modi. Started in 2003, it was conceived as an event to bring investments to Gujarat and to make the state the premier investment destination in the country by showcasing its business-friendly environment, transparent regulatory and administrative regimes and excellent infrastructure.

Gujarat accounts for about 6 % of the land mass of India and approximately 5 % of its population. However, the state now accounts for 7.6 % of the country's GDP, about 10 % of national employment and a whopping 22 % of Indian exports. Between 2001 and 2012, the state had an averaged annual GDP growth rate of 10 %, which was consistently well above the country growth rates. In 1991, it was an electricity deficient state. Through good planning and astute management, it now has a power surplus, with all its villages electrified. Viewed from any direction, Gujarat has made commendable progress over the past 15 years.

Geographically Gujarat is located in a very arid climate, with very few perennial rivers. Not surprisingly and historically, water scarcity has been a very serious issue in the state. Prior to 2001, during years of droughts, severe water crises often led to social tensions and even some years to riots in areas which severely suffered from lack of water. Migrations in search of water, and thus for survival of both humans and animals, were a common phenomenon during the dry summers.

With the construction of major water development projects like the Sardar Sarovar and completion of a Statewide Water Grid, water scarcities have become events of the past.

Because of the importance of water to social and economic development of Gujarat, and standard of living and quality of life of its people, the Government decided to hold one of the seven thematic seminars for the 2015 Vibrant Gujarat on water, climate change and sustainable development.

Gujarat has managed to ensure reliable water supply available for its people, both now and over the foreseeable future. However, with population steadily increasing because of natural causes and migration to the state due to steady and progressive economic development and employment generation as well as rapid increase in human and economic activities, we have to run faster to stand in the same place in terms of ensuring water security. By 2050, per capita freshwater availability in the state will be significantly less than what it is at present. Thus, we have to significantly improve water use efficiencies for all our water-related activities. Accordingly, we are constantly searching for the latest scientific, technological and management advances and then adopt them to improve our water management processes and practices. The main reason for our success has been the formulation of state-of-the-art future-oriented plans and their timely execution.

Our planning and their execution have been consistently underpinned by responsible and proactive political leadership, clear, unambiguous and transparent administrative measures, relentless pursuit of economic growth, and perpetual quest for excellence to ensure better quality of life of our people. All these have contributed to steady improvements in the standard of living of our citizens. The net result, I am pleased to note, has been a true success story of self-reliance and resilience in all water-related areas, creation of consistent growth opportunities and continuing good governance.

While I am pleased with the situations of the past, present and the near future, the water issue has been made more and more complex and complicated due to the uncertainties posed by the expected climate change which is likely to affect nearly all our social and economic activities in somewhat unpredictable ways and rates. This is one of the main reasons why we chose to convene a special seminar during the last Vibrant Gujarat on the interlinkages between water, climate change and sustainable development.

Climate change has the potential to change perceptibly how we plan and manage rural and urban settlements, agricultural production, food availability and distribution, energy generation and use practices, manufacturing and transportation patterns and a host of other development activities. In the area of water, it could alter precipitation patterns and their spatial distribution over time, river flow regimes, groundwater availability and recharge patterns, and variations in water quality.

Aware of the potential impacts of climate change, we established the first Climate Change Department on human development and ecosystem conservation, in the year 2009, in any Indian states.

For our seminar we invited Prof. Asit K. Biswas, who is the Co-founder of the Third World Centre for Water Management in Mexico, to plan a truly world class seminar. He carefully selected and then specially invited the world's leading academics, heads of international organizations and captains of major industries to share their views on how best water, climate change and sustainable development be most successfully managed in Gujarat and the rest of the world. The result was a truly outstanding seminar which brought many new and innovative ideas to the fore. We are analysing them seriously for policy planning and possible implementation.

Because of the outstanding nature and results of this seminar, we decided to put together the best papers that were presented in Gandhinagar. This is the first time during the entire history of Vibrant Gujarat, the results of one of its thematic seminars are being published as a definitive and authoritative book by a major international publisher so that not only Indians but also interested people from all over the world will benefit from the knowledge and ideas that were generated in Gandhinagar.

Finally, let me take this opportunity to express my personal appreciation to Dr. Rajiv K. Gupta and Prof. Biswas for arranging this authoritative and absorbing seminar, and Prof. Biswas and Dr. Tortajada for editing the resulting text. I am confident that the people of Gujarat as well as others from outside the state will benefit significantly from the contents of this book.

Anandiben Patel

Preface

Individually, each of the three issues of this book, water security, climate change and sustainable development, is a difficult, complex and somewhat challenging subject. Academics and policy-makers may often differ with each other as to even the definitions of each of the topics, let alone their implications and ramifications. Thus, and not surprisingly, when these three topics are combined together, their complexities, uncertainties and intricacies multiply by several orders of magnitude.

The fact is, in the real world, irrespective of how each of these three issues are defined, or how they interrelate and interact with each other with numerous known and unknown feedback loops, from a development-related policy perspective they have to be viewed, considered and analysed together. The danger is if each of these three topics is considered independently for formulation of appropriate policy responses, they may have negative implications on the other two issues. This could, for the most part, ensure that the impacts of the pursued policies on the society and the environment as a whole would most likely be sub-optimal, and in many cases even negative over the medium to long terms.

Water security, climate change and sustainable development are closely interrelated and, though difficult and challenging it may be to consider them together, we really have no other choice. Thus, this is a concerted attempt to consider them together from multi-sectoral, multidisciplinary, and multi-issues perspectives.

The book is based on the papers that were presented during an international seminar on the topic during the Vibrant Gujarat Investors' Summit, in Gandhinagar, in January 2015. All the authors were specifically invited to outline their views and thinking on these related issues. Following intensive peer reviews, only the best papers presented were selected. These were then modified by the authors in line with the reviewers' comments. The Government of Gujarat, through its Water Supply and Sewerage Board and Gujarat and Narmada Valley Fertilizers and Chemicals Limited was the main host and organizer of the Seminar with the Third World Centre for Water Management as the main knowledge partner.

Gujarat accounts for 6 % of India's land mass and around 5 % of its population. The state now contributes to 7.6 % of India's GDP, and about 10 % national employment, but a stunning 22 % of the country's total exports. Between 2001 and 2012, Gujarat had an average annual growth rate of 10 %, almost consistently well above India's GDP growth rates. Furthermore, in 1991, Gujarat was a seriously electricity deficient state. However, it is now a power surplus state. All its 18,000 villages have been electrified.

In terms of water security, Gujarat has not been so fortunate. It is situated in a very arid climate within the Asian monsoon belt. Historically, people of the state have suffered serious water scarcity in non-monsoon months. During drought periods, which have been quite frequent, the people have suffered from serious water shortages for drinking, as well as for agricultural and livestock uses. Migrations in search of water have been a recurrent phenomenon during dry summer periods, when social tensions have been quite prevalent, including even water-related riots.

Fortunately, with good water governance practices, and accelerated construction of a state-wide drinking water grid, most of the ravages of water shortages and social tensions of the past have now been basically eliminated.

The socio-economic and water-related developments of the recent past, laudable though they have been, do not guarantee similar levels of continued progress in the future. This is partly because the water regimes of the past appear to be changing, rapidly changing urbanization and development patterns in the state, and equally rapid evolution of global Indian conditions.

Consider climate. Globally, driven by continued economic and population growth rates, anthropogenic emissions are now higher than ever. Annual emission growth rates up to the year 2000 were 1.3 %. Since then, this has increased to 2.2 %. Such developments are changing climate patterns all over the world, including changes in inter-annual and intra-annual precipitations in different parts of the world. This, in turn, is impacting on river flows and groundwater recharge potentials, and thus water availability and use patterns which will undoubtedly impact upon the future water security conditions of the world.

Gujarat has not been immune to these changes. Current data indicate that the frequency of hot days is showing a gradually increasing trend. Furthermore, frequency of cold days seems to be decreasing. Ahmedabad, the largest city in Gujarat, with already over 6 million people, has had serious heat waves in the recent past. In May 2010, temperatures soared to 46.8 °C, which led to heat stress resulting in serious health problems, including deaths, of its citizens. Many have predicted that the temperature may increase by an additional 2 °C which would have serious health, economic and social implications. It is the first Indian city which formulated a heat action plan in 2013.

Climate change may also bring a rise in seawater levels. The average annual rise in mean sea level in Gujarat is estimated to be around 1.3 mm. This is important for the state since it has 1,600 km of coastline, the longest in any Indian state. Rising sea level is likely to increase salinity of groundwater which would affect both humans and ecosystems. A one-metre sea level rise is estimated to affect 14,149 km^2 of area, about 6 % of its coastal population. Higher sea surges in the future may further exacerbate water security problems.

There is no question that water security has to be an integral and essential component of Gujarat's future sustainable development plans. Climate change and subsequent erratic and unpredictable rainfall patterns will affect river flow regimes of the past and groundwater recharge rates. Since a reliable and assured water supply of appropriate quantities and qualities is an essential prerequisite for sustainable development, the interplay and interactions between water, climate change and sustainable development will dictate the future of Gujarat's socio-economic development, as it will in all other Indian states and other countries.

The situation will be especially difficult and complicated in Gujarat since people during the post-2000 period have got used to steadily improving water availability as well as standard of living. They expect the future to be a continuation of the same. It will be a challenging task to meet the people's expectations and aspirations with all the uncertainties imposed on by climate change and rapidly evolving Indian and global conditions.

Fortunately, water is a renewable resource, and with good and efficient management, it can be used and reused numerous times. In addition, in the coming years, there will most certainly be tremendous advances in science and technology as well as management practices to store, use and reuse water. The starting point for solutions to ensure global water security will have to come from strong and effective institutions, political will, appropriate pricing for all types of water uses, good education for all water uses and continuous scientific, technological and management innovations.

We are convinced that the world's water problems are solvable, even after considering the uncertain implications of climate change over space and time.

The contributors to this book come from different disciplines. They represent different sectors like academia, government, business, international organizations and media. All the contributors are leading figures in their fields. We are convinced that this book will make significant contributions to the ongoing global debates and discussions on how we can best link water security, climate change and sustainable development to ensure a bright global future.

And finally we would like to pay compliments to Honourable Chief Minister of Gujarat Smt. Anandiben Patel who not only graced the occasion but also found time from her extraordinarily busy schedule to meet the experts who attended this international Seminar. Due to dedication and support of her team of officers headed by Dr. Rajiv K. Gupta IAS we could organize such a high quality event. We would also like to express our appreciation to Ms. Thania Gomez of the Third World Centre for Water Management, Mexico, for all her work to finalize this book.

Asit K. Biswas
Cecilia Tortajada

Contents

Water Security, Climate Change and Sustainable Development: An Introduction

Asit K. Biswas and Cecilia Tortajada

Abstract The global population is estimated to increase from 7.3 billion at present to about 11.2 billion by 2100. Providing good quality of life for the existing hundreds of millions of people living in poverty and for the additional 3.9 billion people will be a challenging task under the best of circumstances. Many parts of the world are already facing serious water problems, both in terms of quality and quantity. Adding climate change scenarios further intensifies the uncertainties and complexities of the global situation by several orders of magnitude. The realities and politics of water security, climate change and sustainable development are likely to be one of the greatest challenges of the twenty-first century. Each country must rise to and meet these challenges successfully since there are no other alternatives. Status quo or incremental progress will not be an effective option.

A major global concern for all the countries and all the people of the world is how to ensure a high rate of economic growth that is both sustainable and equitable so that hundreds of millions of people who are now living in poverty can have a significantly improved standard of living and quality of life. Equally, the middle class can at the very least maintain their current lifestyles but preferably improve them progressively over time. This has to be achieved with full recognition of the fact that the world population is estimated to increase to 9.7 billion in 2050 and 11.2 billion by 2100, compared to around 7.3 billion at present.

Achieving this goal will not be an easy task because the world will have to eradicate poverty of the existing population and concurrently will have to cater to some 3.9 billion additional people during the next 85 years. Virtually all of these extra people will be in developing countries where lifestyles and standard of living

A.K. Biswas (✉)
Lee Kuan Yew School of Public Policy, National University of Singapore, Singapore, Republic of Singapore
e-mail: akbiswas@cablevision.net.mx

C. Tortajada
Institute of Water Policy, Lee Kuan Yew School of Public Policy, National University of Singapore, Singapore, Republic of Singapore
e-mail: cecilia.tortajada@gmail.com

A.K. Biswas and C. Tortajada (eds.), *Water Security, Climate Change and Sustainable Development*, Water Resources Development and Management, DOI 10.1007/978-981-287-976-9_1

of a majority of the people need to be significantly improved. It is currently estimated that more than 800 million people (14 % of the global population) live in extreme poverty which is defined as having an income of less than $1.25 per day. It should be noted that these global figures by the international organizations, though extensively quoted and mostly taken at face value all over the world, are highly unreliable. In all probability, the numbers are significantly higher than what is now commonly believed.

Unfortunately the growth rates of industrialized countries have been somewhat anaemic in recent years. In addition, China's growth rates, which have been the envy of the world over the past three decades and have been responsible for moving hundreds of millions of the Chinese out of poverty, are now showing signs of trouble and declining from the earlier heady rates. This could be a serious development concern since the country has been an important engine for driving global economic growth in recent decades.

Water is an important component of sustainable development. Not only drinking water is essential for human survival but also it is needed for all types of human activities and endeavours, inter alia, food production, energy generation, resources extraction, industrial development, commercial activities, ecosystem preservation and host of other uses.

Water and energy are two resources which are needed for conducting literally all types of human activities. In addition, they are closely interrelated. Water cannot be produced and used without significant amount of energy, and energy cannot be generated without substantial quantities of water. Thus, however sustainable development is defined or assessed, it cannot be achieved without water and energy securities. Thus, ensuring sustainable development for the world means adequate quantities of water of appropriate qualities are available on a reliable basis over short-, medium or long terms. With the current unsatisfactory levels of water management nearly all over the world, this is unlikely to be possible.

During the Second World Water Forum, held in The Hague, in March 2000, the Ministerial Declaration noted the importance of water security in terms of "ensuring that freshwater, coastal and related ecosystems are protected and improved, that sustainable development and political stability are promoted, that every person has access to enough safe water at an affordable cost to lead a healthy and productive life, and that the vulnerable are protected from the risks of water-related hazards."

Water security of course includes all these issues but it also incorporates many others which the Ministers did not even consider. For example, one important issue that was ignored has been becoming an increasingly serious business risk over the past decades. If fact, the World Economic Forum perception survey of business leaders indicated that water currently is accepted to be the number one business risk in terms of impacts. Lack of adequate quantities of appropriate qualities of water on a reliable basis because of consistent poor governance over the decades has made it a serious risk for many types of business.

If the issue of climate change is added to water security and sustainable development, the problems become infinitely more complex and degrees of risks and uncertainties consequently increase by several orders of magnitudes. The world's water supply is dependent on the rainfall of the past and present which is stored in the ground, lakes and rivers and also retained by man-made structures like dams and barrages. If the rainfall and distribution patterns change in the future, both over space and time, it will make ensuring water security a more complex, formidable and uncertain task. If water security cannot be assured, sustainable development will remain a mirage.

Water management and sustainable development will be affected very significantly because of future climate changes through a variety of pathways, only some of which are known at present and can be predicted with some degree of certainty. There are also likely to be other impacts which are currently unexpected and uncertain and thus are not being considered for any planning or policy-making purposes in any serious and consistent manner.

Climate change will have impacts on various aspects of water management and sustainable development in many different ways. The extent of the impacts on the people, economy and ecosystems will depend on a multitude of factors, and how each of these factors are managed, as well as the interactions between them. Among these factors will be the rates, magnitudes and distribution of the changes, institutional capacities to anticipate and manage the anticipated changes; preparedness of the governments, individuals and the civil society organizations to ameliorate them, advances in science and technology, and rates of adoptions of the new scientific, technological and managerial breakthroughs, as well as social, economic and political perceptions and societal and political willingness to manage the changes.

There are a few countries which are already trying regularly to formulate, update and implement good future-oriented plans which are underpinned by responsible and proactive political and administrative support. These countries will do significantly better than those which are not making comprehensive plans and depending upon ad hoc measures to counteract the potential impacts of climate change as and when they do occur.

Effective future-oriented plans to ameliorate the adverse impacts of climate change should include formulation of good long-term plans which are implementable. It should consider efficient and equitable water management that should take into account the future water needs of the country, both in terms of quantity and quality, and how they could be affected by climate change. Only planning will not be enough: it must be implemented in a timely and cost-effective manner.

This planning has to be conducted within a framework of relentless pursuit of sustainable development that would steadily contribute to steady improvements in economic and social conditions of the people and also meet their future expectations. This will be an exceedingly complex and difficult task which only a select few countries will be able to do properly in cost-effective and socially acceptable ways.

A good example is Singapore, a low-lying island where much of its land lies within 15 m above the mean sea level. Some 30 % of its land area is within 5 m of the mean sea level. Thus, for a country like Singapore, the most immediate threat due to climate change is sea level rise.

In order to reduce its vulnerability from coastal erosion, nearly 80 % of the island state's coasts now have hard walls or stone embankments. The rest are natural areas like beaches or mangroves.

In 1991, Singapore made a policy decision that all newly reclaimed land must be built at least 1.25 m above the highest ever recorded tidal level. As the threat of climate change increased and the scientific evidence indicated that sea level rise may be higher than what was expected earlier, in 2011, the country decided to raise the minimum level of the newly reclaimed land at least by an additional metre. This could ensure that the country could be safeguarded from the long-term impacts of higher anticipated sea level rise. Because of such continuous long term planning and its execution, the country is likely to fare much better in adapting to the new conditions due to climate change compared to all its neighbours.

The social, economic, environmental and political impacts of climate change, both over space and time, are complex interacting issues with numerous known and unknown feedbacks loops which are still not fully understood at present. Lack of proper scientific understanding of how various physical, social, economic and environmental forces may interact with each other, as well as absence of reliable data and information, are only two factors which currently prevent most countries to make reasonably reliable and actionable predictions of future multidimensional, multi-sectoral and steadily evolving impacts. These uncertainties further contribute to taking timely planning and investment decisions in a cost-effective and socially-acceptable manner. These are truly most challenging tasks under the best of circumstances.

At our present state of knowledge there are a number of important uncertainties in predicting the impacts of climate change on the water sector. These include, but are not necessarily limited to, future global emission scenarios, analysing them for various greenhouse gas emissions, predicting how these are likely to affect future global, regional and local rainfall patterns over time and space, interpreting their overall impacts on the hydrologic cycle at different scales including river flow regimes, assessing the types of scientific and technological breakthroughs expected that would allow us to better understand, predict and ameliorate the various issues associated with climate change and evaluating their impacts. These are likely to be mammoth tasks.

At our present state of knowledge, it is not possible to predict even how the annual average temperatures in various regions of the world may change over time, let alone the rainfall distribution patterns. In addition, even if such average global and regional changes in precipitation patterns could be predicted with some acceptable degree of reliability, they will be of very limited use for water sector planning and management. It would be essential to obtain information at river basin and sub-basin levels, which are often units of planning: country-level information will be of very limited value. The reliabilities of the current generation of hydrologic models have to improve by several orders of magnitudes before they can be used for planning and decision-making purposes with any degree of confidence.

The problem becomes even more complex for the water sector where it is essential to have not only reasonably reliable information on the distribution of rainfall at the river basin and sub-basin levels, but also, more importantly reliable information on extreme rainfall events which would contribute to heavy floods and prolonged droughts. It is a fact of life that water structures are not designed on the basis of average rainfalls but primarily on extremes of rainfalls.

The magnitudes of uncertainties increase several fold when attempts are made to forecast extreme rainfall events on the basis of increases in average annual temperatures, and even more when rainfalls have to be translated into river flows. The peak river flows not only depend on high rainfalls, though it is a major factor, but also on many other factors, including land use changes and efficiencies and capacities of institutions responsible for water planning and management.

For successful and intelligent planning of the water sector within the framework of sustainable development, knowledge needs to advance much more than what is available at present. Meeting these challenges does not depend exclusively on advances in climatological-hydrologic modelling. Policies for adaptation and strategies for cost-effective mitigation measures have to be formulated on the basis of what are likely to be potential impacts. The forecasts of these impacts have to be periodically fine-tuned and considered for implementation according to changing societal and political expectations and also availability of additional knowledge, data and information.

Even more challenging will be the politics of policy-making and implementing which will require a quantum jump in existing planning and decision-making practices and processes. The politics of water security, climate change and sustainable development is likely to be one of the greatest challenges of the twenty-first century. How the individual societies will navigate these challenges in the coming decades will most likely determine the levels of their relative successes.

These will be immensely difficult, complicated and challenging tasks. However, countries must rise to meet these challenges successfully since there are no other real alternatives. Status quo or incremental progress will not be an effective long-term option.

Climate Change, Resource Efficiency and Sustainability

Peter Brabeck-Letmathe

Abstract Nestlé has developed numerous strategies along the years to address climate change, resource efficiency and other long-term challenges within a sustainability framework. These strategies have given the company the opportunity to innovate towards more environmentally sustainable and science-based products and services. Nestlé has successfully and continually improved resource use efficiency in water, energy and materials in a perspective of reducing exposure to risks of disruption. The Group championed the use of clean and renewable energy, as well as reducing waste at all stages of production and other parts of the value chain it can influence. The implementation of the sustainability framework has, among other things, resulted in significant reduction of greenhouse gas emissions.

1 Introduction

At Nestlé, we take the risks of climate change very seriously. This is not only because of Nestlé's strong commitment to its responsibilities as a global enterprise, but also because of the risks and impact climate change could have on Nestlé's business in the future. Rising global temperatures will cause sea levels to rise and alter local climate conditions, which may also affect forests, agricultural productivity, water supplies, human health, animals, and many types of ecosystems (Olesen and Bindi 2002; McMichael et al. 2006; Fischer et al. 2007; Allen et al. 2010).

A sustainable development strategy framework led by Nestlé relies on systemic coordinated solutions. We address possible solutions in several ways. One of them is resource efficiency. We believe that the efficient and scrupulous use of natural resources—including energy, water and materials, and the use of more renewable and clean energy—lie at the heart of climate change mitigation. Some research carried out on this topic has provided us with supportive evidence, in-depth analysis, and even inspiration on feasible methods (Barrett and Scott 2011; Abdullah et al. 2014).

P. Brabeck-Letmathe (✉)
Board of Nestlé SA, Vevey, Switzerland
e-mail: Karin.Janssens@nestle.com

A.K. Biswas and C. Tortajada (eds.), *Water Security, Climate Change and Sustainable Development*, Water Resources Development and Management, DOI 10.1007/978-981-287-976-9_2

By recognizing the importance of resource efficiency for mitigating climate change and maintaining sustainability, Nestlé has moved towards resource efficient, renewable energy-oriented and 'zero waste' production. In the past decade, our efforts to improve resource efficiency, utilize clean and renewable energy and reduce waste have contributed to reducing greenhouse gases (GHG), strengthening nature's resilience and facilitating adaptation to climate change.

2 Nestlé's Strategies and Perspectives as the World's Leading Nutrition, Health, and Wellness Company

2.1 Nestlé's Strategies and Perspectives on Resource Efficiency

As the world's leading Nutrition, Health, and Wellness company, we use resources as efficiently as possible in our food manufacturing process. At this stage we can directly manage and improve our resource performance. Resource efficiency is aimed both at mitigation and adaptation/increased resilience.

Our strategies on resource efficiency focus on the following aspects:

At Nestlé, in support of the ISO 14001 certification that focuses on environmental management, we use the most efficient technologies and apply the best practices to optimize our manufacturing process, reduce energy and water consumption, minimize waste generation, utilize sustainably-managed renewable energy sources, recover value from by-products and control, and eliminate emissions including greenhouse gases.

In the packaging process, we optimize the weight and volume of our packaging and lead the development and use of materials from sustainably-managed renewable resources, while considering packaging and product performance requirements. Furthermore, we support initiatives to recycle or recover energy from used packaging, and use recycled materials when appropriate and where there is an environmental benefit.

At the distribution stage, we optimize distribution networks and route planning across all our operations, and also explore opportunities to improve transportation efficiency and reduce air emissions. We have expanded our driver training both from a safety and environmental efficiency perspective, and use telematics and the latest technology on our vehicles where practical, recommending the same to our suppliers (Nestec Ltd. 2013).

To achieve sustainable development, besides the continuous improvement of resource efficiency in Nestlé's food production process, we are exploring opportunities to improve resource efficiency in the upstream part of our supply chain, such as in crop cultivating and harvesting and pre-processing processes. We are also doing this in the downstream part of our products' value chain in the retailing and food consumption processes, among other things. This is due to the fact that energy consumption in the manufacturing process of our products actually only accounts for a relatively small proportion of energy used in the entire food value chain from farmers to consumers (Nestlé CSV Report 2013, p. 233). Upstream, "our work in agriculture involves cross-sector partnerships, where we marry our technical expertise and research in crop optimization with our partners' local community knowledge. In the process, we help farmers increase their productivity and help ensure access to high quality crops for our business." (Brabeck-Letmathe 2006). Downstream, on one side, Nestlé brings together specialists and researchers to innovate and optimize the design of its products in order to help consumers achieve higher resource efficiency during the consumption of products. Nestlé is also dedicated to train its suppliers, consumers and employees with the skills of reducing waste and using resource more efficiently.

Resource efficiency alone, while absolutely essential, is not the only solution. It has been noted that technological improvements to increase the efficiency of a resource could lead to an increased consumption of this resource in the industry (Jevons 1957). Twentieth-century economic growth theory has also shown that technological change is the main cause of increased production and consumption (Alcott 2005). At Nestlé, while improving resource efficiency, we are also pursuing total resource consumption reduction in our food manufacturing and delivering processes. This certainly will be a bigger challenge for us since our production volume is increasing steadily.

Nestlé has an integrated and holistic approach towards climate change and resource efficiency because energy, greenhouse gas emission, water and materials are all interlinked in this context. Considering GHG emissions in isolation may have a detrimental impact on other environmental aspects, such as water. Water and energy are interactive—improving the efficiency of one will have an impact on another. Although recycling is efficiently helping to reduce resource consumption and GHG emissions, recycled materials do not always have an environmental benefit over virgin material. With a deep understanding of this energy-water-material-emission relationship, and by taking steps to manage these resources in a more holistic and strategic manner, we are able to improve resource efficiency in an integrated framework, rather than the efficiency of any specific resource.

We would also like to point out that the criteria of sustainability are dynamically developing and changing over time. In order to achieve sustainable development, we carefully monitor, evaluate and communicate regulatory developments so that they are timely reflected in our corporation strategies. We are also proactively engaging with regulators and other relevant stakeholders long-term to foster environmentally efficient and effective laws and regulations, and favour the harmonization of environmental laws, regulations and standards in order to define, implement and evaluate solutions to the complex environmental challenges we are facing (Nestec Ltd. 2013).

We will continue to implement these strategies of resource efficiency and environmental sustainability so that Nestlé's products are not only tastier and healthier, but also better for the environment along their value chain. We believe in the future, in the highly competitive food and beverage market, more environmentally friendly and sustainable food products are more attractive to consumers and will gain further loyalty from them. Nestlé's efforts to promote environmental sustainability also means delighting its consumers by giving them more reasons to trust our company, enjoy our products, and live up to employees' and external stakeholders' expectations about our environmental responsibility and practices.

2.2 *Creating Shared Value*

Although climate change brings challenges to our business, at Nestlé we see climate change as an opportunity to innovate towards more environmentally sustainable and science-based products and services, rather than perceive it as a threat. Through continuous innovation we are improving efficiency at every stage of our food production process, reducing our material and energy consumption, and in turn, achieving lower costs for our products, which are giving us more competitive advantage in the food market. We believe that over the long-term, we create value for our shareholders and for society as a whole. We call this 'Creating Shared Value' (CSV).

As Kramer (2006) said: "CSV is a very different approach to Corporate Social Responsibility (CSR) because it is not focused on meeting a set of standard external criteria, or philanthropy. Rather, we are talking about creating social and environmental benefit as a part of making a company competitive over the long-term. … Business can help societies progress and all sectors can help business improve and flourish." Shared value is not about 'sharing' the value already created by firms as a redistribution approach, but about expanding the total pool of economic and social value—a new way to achieve economic success. It is not on the margin of what companies do, but at the centre (Porter and Kramer 2011).

CSV is at the heart of Nestlé's development strategy that goes beyond corporate responsibility to obtain a win-win for our shareholders and society. In the lowest tier of CSV, we ensure compliance with national laws and relevant conventions, as well as with Nestlé's own regulations. Beyond compliance in the second tier of CSV, Nestlé's business operation is based on sustainability to ensure that our activities today preserve the environment for future generations. While at the top of the CSV hierarchy, this goes further beyond both compliance and sustainability. Nestlé believes that businesses can finally create value for both its shareholders and society through its activities when they have far-sighted visions, and follow environmentally sustainable principles in the long-term.

Nestlé's CSV strategy is not only recognized within the Nestlé family but also by academic researchers. Porter (2011), a leading researcher in business strategy and a professor at Harvard Business School, commented on our 'Ecolaboration' initiative by Nespresso, a programme built on Nestlé's approach to CSV: "Nespresso's innovative programme, which helps farmers achieve higher prices, better yields, and greater environmental performance and sustainability, is not driven by charity but by creating value. Nespresso will benefit strategically from the quality improvements that farmers achieve and coffee supplies that will be far more sustainable over time. It is this alignment between corporations and social challenges, not a mind-set of separation and trade-offs, which is the key to both economic and social progress."

3 Nestlé's Efforts to Improve Resource Efficiency and Approach Sustainability

3.1 Improving Energy Efficiency

Over the past decade, Nestlé has made considerable efforts to innovate techniques and improve energy use efficiency in its factories, plants and even in its offices. Our goal is to become the most efficient energy user among food manufacturers. We would like to share some of our achievements:

- Nestlé Waters has invested CHF 51.8 million to build a new factory in Buxton, United Kingdom, which is one of the most innovative and efficient bottling facilities in the world. This new bottling and warehousing plant has enabled our water business to significantly lower its energy consumption and to cut packaging by an average of 25 % across the *Buxton* and *Pure Life* ranges (Nestlé CSV Report 2013, p. 216).
- In 2013, Nestlé optimized more than ten distribution networks globally to cut gasoline consumption and CO_2 emissions. The redesign makes better use of space in our vehicles, avoiding unnecessary miles and using more efficient modes of transport (Nestlé CSV Report 2013, p. 248).
- Nestlé's new employee centre in the state of Chiapas in Mexico became the first building in Latin America to be certified with the prestigious LEED Platinum standard. We have invested more than CHF 800,000 in the building. It can save about 50 % of energy and 80 % of water use compared to a conventional building (Nestlé CSV Report 2012, p. 175).

In 2012, 850 energy-saving projects were launched, comprising a total investment of about CHF 82 million. Annual energy savings of about two million giga joules and a reduction of approximately 173,000 tonnes of CO_2 equivalent (CO_2eq) are expected from these projects (Nestlé CSV Report 2012, p. 172).

We aim to continue to pursue energy efficiency in our factories. Nestlé has globally achieved total energy consumption reduction by 22.6 % per tonne of product since 2005, as summarized in Table 1 (Nestlé CSV Report 2013, p. 233).

At Nestlé, we not only focus on improving energy efficiency in our manufacturing processes, but we also try to help consumers improve their energy use efficiency during the consumption process of our products. Guided by this strategy, Nestlé has brought together specialists from internal research and development teams and manufacturers to design more innovative, high performing and energy-efficient coffee machines. Our innovative designs have helped our consumers to improve water and energy efficiency and save total resource use during

Table 1 2013 energy use per product category

Product category	Energy (GJ/tonne)		
	2005	2013	Variation (%)
Powdered and liquid beverage	17.8	10.03	−41.3
Water	0.39	0.24	−38.8
Milk products and ice cream	4.93	3.09	−37.3
Nutrition and healthcare	10.7	7.35	−31.3
Prepared dishes and cooking aids	3.93	3.5	−10.9
Confectionery	5.89	4.45	−24.3
Pet care	1.81	1.82	0.5
Total Group	2.42	1.87	−22.6

the consumption process. For example, a standby function has been introduced in the CS professional machine range for our business customers in Europe, which can be activated automatically after a time delay chosen by the customer. During standby mode, the energy consumption of the CS220 machine is less than 0.5 W. This is over 20 times lower than when the machine is switched on and in 'ready to use' mode (Nespresso).

Through continuous improvement of energy use efficiency, we are able to keep the total consumption of energy in our food production almost unchanged in comparison to 2003, even though our production volume has had a steady increase over the past decade. Between 2003 and 2013, our energy consumption increased only by 3.5 %, while our production volume increased by 53 % (Nestlé CSV Report 2013, p. 232). Considering Nestlé's total energy consumption of 97.7 billion mega joules on-site in 2013, this achievement brings significant benefits to our business operation, and to the global environment.

The "energy use efficiency" for Nestlé is calculated based on annual sales as the total energy use per dollar of sales, which was 0.98 mj/dollar in 2013. In comparison with other major food and beverage companies, such as General Mills, Nestlé has relatively higher energy consumption per dollar of sales, implying it could further improve its energy use efficiency, although energy consumption is also related to the products portfolio.

3.2 Greenhouse Gas (GHG) Emission Reduction: Cleaner and/or Renewable Energy

Besides improving energy use efficiency, our approach to air emission reduction includes switching to sustainably-managed clean fuels and renewable resources (Nestlé CSV Report 2013, p. 265). Over the past ten years, enormous efforts have been made towards the increase of renewable energy percentage in Nestlé's factories and plants worldwide. Below are some examples of our recent renewable energy initiatives:

- In Italy, Nestlé Waters replaced its oil-powered boiler with a biomass boiler, producing 100 % renewable energy at the Pejo plant. Wood production and waste wood chips from within 40 km of the plant are being used to operate the boiler. The Pejo plant has a long-term goal of providing enough energy to connect nearby towns to a network grid that utilizes renewable energy from the biomass boiler (Nestlé Waters CSV Report 2011, p. 57).
- In Mexico, 85 % of the total electricity consumed by our factories is now supplied by wind power following a power purchase agreement with the Mexican wind-turbine company CISA-GAMESA, which will reduce air emissions, including GHGs, by an estimated 124,000 tonnes of CO_2eq annually. This is comparable to taking 39,000 small cars off the road. Nestlé is the first food company in Mexico that consumes nearly all the electrical energy needed for its manufacturing operations from a renewable source (Nestlé CSV Report 2012, p. 193).
- In Chile, Nestlé has already invested CHF 5 million to install and convert factory equipment in its Osorno, Cancura and Llanquihue milk factories by switching from coal and fuel oil to natural gas. It is projected that 30 % of GHG emissions across the three sites will be reduced and an annual cost of CHF 1 million at Llanquihue will be saved in the future (Nestlé CSV Report 2013, p. 235).

In 2013, the share of renewable energy consumption reached 13.3 % of the total energy consumption on-site at our factories. This ratio was the result of an increase from 12.2 % in 2012 (Nestlé CSV Report 2013, p. 265) and 11.6 % in 2011 (Nestlé CSV Report 2012, p. 265). By investing in renewable sources, improving energy efficiency and switching to cleaner fuels, Nestlé has reduced a total of direct GHG emissions by 45 % since 2003 per tonne of product. It has also cut CO_2eq by 15 % to four million tonnes, while increasing production by 56 % (Nestlé CSV Report 2013, p. 20). Although indirect GHG emissions, which come from purchased energy increased by 23 % to 3.8 million tonnes since 2007, a reduction of 3 % of indirect GHG emission per tonne of product was achieved (Nestlé CSV Report 2013, p. 266).

According to the sustainability report by PepsiCo in 2013, its total direct and indirect GHG emissions for legacy operations in 2013 were a little bit more than four million tonnes of CO_2eq, which represents a decrease of 2 % versus 2012 (PepsiCo Sustainability Report 2013, p. 39). The GHG emissions levels were constant from 2008 to 2012 (PepsiCo Sustainability Report 2011/2012, p. 24).

The General Mills Global Responsibility 2014 Report shows that General Mills gradually reduced its GHG emissions from 1.04 to 0.97 million metric tonnes in the past five years (General Mills Global Responsibility 2014, p. 40).

The American food company Tyson has highlighted in its Sustainability Report (2013) that its absolute GHG emissions were 5.22 million metric tonnes in 2012, which is slightly higher than in 2011 and 7.4 % lower than the level in 2010.

The "so-called" specific CO_2eq emission, defined as the total GHG emissions (both direct and indirect) per dollar of sales, was 0.08 kg/dollar for Nestlé in 2013. Compared to other food and beverage companies, such as General Mills, PepsiCo

Table 2 Specific CO_2eq emission comparison

Company	Item	2010	2011	2012	2013
General Mills	Sales (billion dollar)	14.64	14.88	16.66	17.77
	CO_2 emissions (million metric tonnes)	1.03	1.01	0.96	0.97
	Specific CO_2 efficiency (kg/dollar)	0.07	0.07	0.06	0.05
Nestlé	Sales (billion CHF)	93.02	83.64	89.72	92.16
	CO_2 emissions (million metric tonnes)	7.12	7.03	7.09	7.81
	Specific CO_2 efficiency (kg/dollar)	0.08	0.07	0.07	0.08
PepsiCo	Sales (billion dollar)	57.84	65.88	65.49	66.42
	CO_2 emissions (million metric tonnes)	4.22	4.22	4.22	4.14
	Specific CO_2 efficiency (kg/dollar)	0.07	0.06	0.06	0.06
Tyson	Sales (billion dollar)	28.4	32.25	33.3	n.a.
	CO_2 emissions (million metric tonnes)	5.64	5.2	5.22	n.a.
	Specific CO_2 efficiency (kg/dollar)	0.2	0.16	0.16	n.a.

Data is extracted from General Mills Annual Report (2013), Nestlé Annual Report (2011, 2013), PepsiCo Annual Report (2011, 2013), 10-K Annual Report 2013—Tyson, 10-K Annual Report 2012—Tyson, General Mills Global Responsibility (2014), PepsiCo Sustainable Report (2011/2012, 2013), Nestlé CSV Report (2011, 2012, 2013), Tyson Foods Sustainability Report (2013/2014), annual average exchange rate is sourced from http://www.ozforex.com.au/

and Tyson, Nestlé has achieved one of the best-in-class in terms of level of specific CO_2eq emissions as shown in Table 2.

Over the past two decades, carbon intensity has been significantly reduced in many countries especially in developing countries such as China and Russia. Figures 1 and 2 show the historical trends of carbon intensity in a few major developed and developing economies in the world (data is extracted from United States Millennium Development Goals database).

3.3 *Improving Water Use Efficiency*

Water use efficiency and sustainability cannot be addressed in isolation, but together with other stakeholders. This is the approach of the international 2030 Water Resources Group, also active in India, a group that I am chairing. But there are also efforts at corporate level for effective water use. We recognize that this critical natural resource must be used efficiently at our factories and distribution centres. In 2013, we approved 38 million water-related investments and implemented 171 programmes in our factories to reduce and reuse water resources, which have saved 3.6 million cubic metres of water (Nestlé CSV Report 2013, p. 183). We have set robust water efficiency targets up to 2015, which continue to drive improvement programmes across all of our operational sites. Some of the programmes that we implemented in 2013 to improve water use efficiency in our plants and factories are listed below, demonstrating our tremendous efforts in this crucial area.

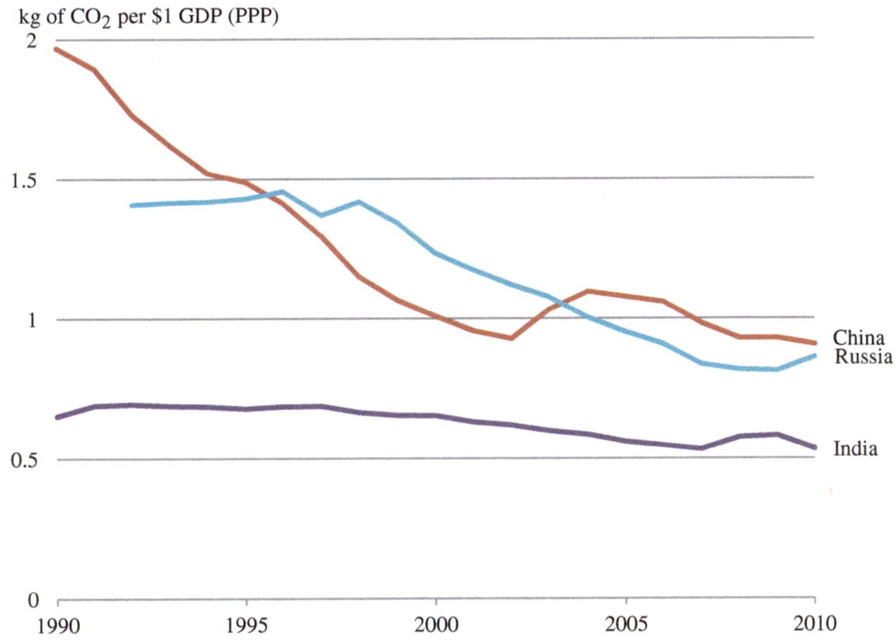

Fig. 1 Carbon intensity of developing economies

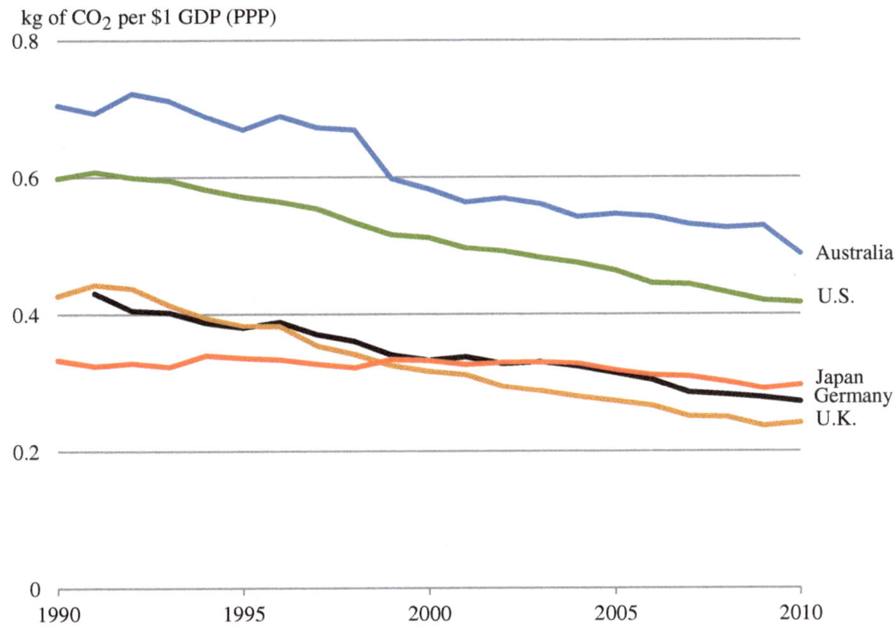

Fig. 2 Carbon intensity of developed economies

Table 3 Water withdrawals per product category (Nestlé CSV Report 2013, p. 183)

Product category	Water withdrawals (m^3/tonne)		
	2005	2013	Variation (%)
Powdered and liquid beverage	22.91	9.85	−57.0
Water	1.96	1.58	−19.4
Milk products and ice cream	8.33	4.77	−42.7
Nutrition and healthcare	19.34	11.02	−43.0
Prepared dishes and cooking aids	7.12	4.22	−40.7
Confectionery	9.22	4.7	−49.0
Pet care	1.44	1.05	−27.1
Total	4.41	2.92	−33.8

- In northern Spain, Nestlé's confectionery factory at La Penilla brought together a multidisciplinary team of employees to identify opportunities of water conservation. With a CHF 1.2 million investment, the regulation of water flow through our milk evaporators was modified and one million cubic metres less water is used every year. In less than 12 months, water use per tonne of product was reduced by 60 % without increasing energy consumption or greenhouse gas emission at the factory (Nestlé CSV Report 2013, p. 182).
- Our recent initiative, Zéro Eau (zero water), has been implemented in water-scarce areas to help improve water availability in the community. In Mexico, Phase One of the project involving the recovery and use of condensation from the milk evaporation process has been completed. Along with various saving initiatives, it delivered annual water savings of 373,300 m^3 in 2013. When the project is completed, the factory will become self-sufficient in water consumption—zero consumption of fresh water (deep well) during normal production (Nestlé CSV Report 2013, p. 183).

In recent years, improvement in water use efficiency is substantial at Nestlé. Between 2005 and 2013, direct water withdrawals in every product category have decreased and an overall reduction per tonne of product of 33.3 % since 2005 was achieved (Nestlé CSV Report 2013, p. 183) (Table 3).

We have also significantly reduced water discharge per tonne of product by 60.1 % since 2003 and total quantity by 37.2 %. We recycled or reused 6.7 million cubic metres of water in our operations during 2013.

Table 4 shows Nestlé's water discharge in total and per tonne product, average quality of water discharge and water reused data from 2010 to 2013.

While improving water use efficiency at our factories and distribution centres, we are also exploring opportunities to improve water use efficiency in our upstream supply chain—in the farmers' production process.

In 2010, a three-year study began which aimed to measure consumptive use of water at farm level, develop best practices and disseminate recommendations to improve water use. Different scenarios of combinations between groundwater and rainfall used for irrigation were made based on climatic data and interviews with

Table 4 The indicators of water discharge (Nestlé CSV Report 2011, 2012, 2013)

Indicator	2003	2008	2009	2010	2011	2012	2013
Total water discharge (million m^3)	145	96.1	91	94	93.9	84.2	91
Total water discharge (m^3 per tonne product)	4.36	2.34	2.22	2	2.08	1.77	1.74
Average quality of water discharge (mg COD/l)	n.a.	95	90.8	78	68.6	93.7	76.2
Water recycled or reused (million m^3)	n.a.	n.a.	n.a.	n.a.	7.8	6.9	6.7

over 300 coffee farmers. As a result, recommendations from the study may result in more than 50 % water savings versus conventional practices (Nestlé CSV Report 2013, p. 175). The farmers who supply to us benefit from the practice and guidance from various Nestlé sustainable studies with help of thousands of Nestlé agronomists (Nestlé CSV Report 2013, p. 227).

Besides helping farmers improve water use efficiency on irrigation in the upstream supply chain, Nestlé works together with local farming associations and non-governmental organizations to build central mills for coffee farmers in order to use water more efficiently during the coffee cherries processing process. In the central mill built by Nespresso and partners in Jardín, Colombia, 63 % of water usage is saved compared to home processing and waste water is 100 % treated thanks to better milling techniques. On average, it saves 27 l of water per farm and potentially helps to increase farmer profitability by 30 % (Project News Report; Nespresso).

In the meantime, downstream the value chain of our products, we are helping consumers to improve water use efficiency and reduce water consumption during their consumption process through innovations. Two of the examples are provided below:

- The newly designed 'eco-mode' function for our *Nescafé Dolce Gusto* Melody coffee machine helps consumers cut water consumption by 25 % when making a 120 ml cup, compared to the first model launched in 2006 (Nestlé CSV Report 2013, p. 199).
- *Maggi* So Juicy chicken recipe mixes enable the chicken to cook in its own juices inside a specially-designed cooking bag. This design aims to help consumers reduce the amount of water and chemicals required for oven cleaning (Nestlé CSV Report 2013, p. 254).

The issues of water and energy are highly interactive and inextricable. Manufacturers use significant amounts of energy to pump, purify, heat and clean the water that flows through their many operations. Equally, energy use frequently requires large amounts of water, such as for cooling in industrial processes. As a result, improving the efficiency use of one has an impact on another.

Nestlé has estimated the water footprint for dairy products produced at Nestlé Colombia's Florencia and Bugalagrande sites. The results demonstrate that in 2012,

Table 5 Water withdrawal efficiency comparison

Company	Item	2010	2011	2012	2013
Danone	Sales (billion Euro)	17	19.3	20.9	21.3
	Water withdrawal (million m^3)	56.25	57.33	58.71	60.63
	Water withdrawal (litre/dollar)	2.49	2.13	2.18	2.14
Nestlé	Sales (billion CHF)	93.02	83.64	89.72	92.16
	Water withdrawal (million m^3)	144	143	138	152
	Water withdrawal (litre/dollar)	1.61	1.51	1.44	1.53

Data is extracted from the Nestlé Annual Report (2011, 2013), Nestlé CSV Report (2011, 2012, 2013), Danone Annual Financial Report (2013), Danone Sustainability Report (2010, 2012), Annual average exchange rate is sourced from http://www.ozforex.com.au/

58 % of the consumed water at the plants is related to the dairy farms, while 35 % is related to electricity and fuel consumption. After actions were taken to improve water use efficiency, the water withdrawal at the Florencia plant was reduced by 44 %, from 4.9 to 2.7 m^3 per tonne of dairy product. During the same period, electricity consumption per tonne of product has also been reduced by 35 % (Nestlé CSV Report 2013, p. 180).

Since 2003, Nestlé has succeeded in reducing the total amount of water in its manufacturing processes. Its total water withdrawal has decreased by 21.2 % despite a 56 % increase in its production volumes between 2003 and 2013 (Nestlé CSV Report 2013, p. 19).

According to the 2013 annual report by the French food company Danone, its total water withdrawal from all sources was 60.63 million cubic metres in 2013, which is 3.3 % more than its 2012 water withdrawal (Danone Annual Financial Report 2013, p. 176).

Table 5 shows a comparison of the water withdrawal efficiency from Danone and Nestlé, based on annual sales value from 2010 to 2013.

3.4 *Improving Materials Use Efficiency*

The packaging of Nestlé products is crucial to guarantee our high quality standards and to prevent food waste. Our strategies on improving packaging material efficiency focus on two dimensions: reducing material demand through innovative and optimized designs, and recycling packaging materials.

In 2012, Nestlé Waters represented around 48 % of Nestlé's overall packaging reductions. Packaging innovation and optimization is particularly important for Nestlé Waters. Through its innovation and optimization programmes, Nestlé Waters has developed a new generation of lightweight bottles, caps and labels, which reduced packaging weight per litre by 17 % across its operations globally between 2007 and 2012 (Nestlé CSV Report 2013, p. 237).

Below are some of our achievements on packaging innovations:

- In Brazil, Nestlé Waters launched a new 0.5L polyethylene terephthalate (PET) bottle in 2012 that is 20 % lighter than the previous version. Today, it is the lightest bottle in the bottled water market in Brazil (Nestlé CSV Report 2013, p. 237).
- In 2010, *Nescafé Gold* launched a new jar in five countries in Europe. Packaging lines were upgraded to minimize collisions, leading to lower breakage rates and reduced noise level. Since then, 650 tonnes of glass per year have been saved. This initiative not only cuts waste during production, but also improves the safety of our employees (Nestlé CSV Report 2013, p. 237).
- In 2013, Nestlé Thailand reviewed the design of its small format *Nestlé Pure Life* and *Minere* PET bottles, which included replacing the 0.5L bottle with a new 0.6L bottle. This resulted in an average weight saving of 10 % for the 0.6L and 1.5L formats. An overall 850 tonnes of PET was saved (Nestlé CSV Report 2013, p. 244).

In 2013, Nestlé saved 66,594 tonnes of packaging material by weight in comparison to 47,000 tonnes in 2012, equivalent to a cost saving of CHF 158.5 million. We have now saved more than half a million tonnes of packaging materials since 1991, and achieved a CHF 1 billion saving milestone in 2012 in packaging costs since 1991. This contributes significantly to the environmental benefits: over the last five years, we have saved over 490,000 tonnes of CO_2 eq thanks to savings in packaging material—that's the equivalent to taking more than 106,000 cars off the road for a year (Nestlé CSV Report 2013, p. 237).

While improving material use efficiency and reducing material demand, the recycling rate increase of materials during the packaging process is also a focus for Nestlé. We use recycled content in our packaging such as paper, cardboard, PET, glass and tinplate while making sure that the safety and quality of the product is not compromised.

- Nespresso has been focusing on recycling of its aluminium capsules for more than 20 years. More than 14,000 dedicated capsule collection points were installed across 27 countries, plus a home collection service is now active in 15 countries. In June 2012, it achieved 76.4 % capacity with its collection systems, which surpassed its 2013 target of 75 % recycling capacity (Nestlé CSV Report 2013, p. 242).
- Nestlé Waters has also improved its packaging material use efficiency by recycling used materials. In 2013, Nestlé Waters created a new 'ReBorn' 0.5L water bottle for its *Arrowhead* water brand, made from 50 % recycled plastic (Nestlé CSV Report 2013, p. 244).

Annually between 2011 and 2013, 38.9 % of materials for packaging were renewable and 27.1 % of recycled materials were used for packaging purposes in Nestlé factories and plants worldwide. Table 6 shows the indicators regarding packaging material use and recycling at Nestlé since 2008.

Table 6 Indicators of material use efficiency (Nestlé CSV Report 2011, 2012, 2013)

Indicator	2008	2009	2010	2011	2012	2013
Materials for packaging purposes (million tonnes)	n.a.	4.2	4.59	4.58	4.8	5.3
Packaging materials against production volume (tonne/tonne)	n.a.	0.1	0.11	0.1	0.11	0.11
Packaging source optimization (1000 tonnes saved)	58.6	59	70.8	39.3	47.1	66.6
Cost saving (million CHF)	n.a.	n.a.	n.a.	65	94	158.5
Renewable packaging materials (% of materials for packaging purposes)	n.a.	n.a.	n.a.	38.9	38.9	38.9
Total % of recycled material in our packaging (% of materials for packaging purposes)	n.a.	n.a.	n.a.	27.1	27.1	27.1

Although recycling used material is an effective measure to reduce GHG emissions and costs, its environmental performance is not always better. Recycled materials do not always have an environmental benefit in comparison to virgin material. For example, in some instances we would need a heavier weight of recycled materials to guarantee our packaging standards. Therefore when we implement recycling strategy at Nestlé, we adopt a holistic and integrated approach to evaluate the environmental benefit, rather than focusing only on one indicator.

3.5 *Zero Waste*

In a world that faces an increasing constraint of natural resources, we recognize the importance of reducing waste and using our resources efficiently. 'Zero waste' for disposal was therefore established as one of the goals in Nestlé's operations, aiming to recover and reuse materials to create additional values—from energy recovery to animal feed. No factory waste will go to landfill or be incinerated without energy being recovered from the process. For example:

- At Nestlé's Fawdon factory near Newcastle in the UK, we piloted an on-site anaerobic digestion system. The facility turns liquid and solid waste into energy and reduces discharge going to the sewer. A full size digester was installed in 2013, which enables us to make better use of waste, and cut the factory's electricity consumption by 8.9 %, resulting in a 3 % reduction in total energy (Nestlé UK).
- Nestlé Malaysia started a project in 2013 to achieve zero waste to landfill in five factories by 2016. Activities include recycling laminates to make table tops and roofing, and using microbial conversion to transform sludge—a by-product of wastewater treatment—into fertilizer (Nestlé CSV Report 2013, p. 258).
- In 22 Nescafé factories worldwide, energy from the coffee grounds resulting from the manufacturing process was recovered as a source of renewable energy.

> Product losses were processed so they can be added to farm animal feed. Other materials such as metals, plastics, paper, cans and cardboard are processed by contractors and traded as commodities. Any remaining material that does not currently have a viable recycling option is sent for incineration, with energy recovery as a first option (Nestlé CSV Report 2013, p. 260).

In line with Nestlé's 'zero waste' strategy, since 2008 we have strongly reduced waste for disposal per tonne of product by 51 % down to 4.9 kg per tonne, and we have cut the total waste for disposal by 37.3 % to 257,000 tonnes at Nestlé.

Between 2011 and 2013 our hazardous waste, such as detergents, oils, fuels and grease, has been reduced from 55,000 to 32,000 tonnes, which accounts for only 1 % of the waste for disposal at Nestlé. The generated hazardous waste is managed and treated in accordance with local regulations and standards worldwide (Nestlé CSV Report 2013, p. 258).

Up to 2013, we have increased our waste and by-product recovery rate to 85 %, meaning that we reused and recycled 85 % of the waste we produced in 2013. Of the waste we disposed of, 86 % went to landfill, 4 % to incineration and 10 % to other methods of disposal (Nestlé CSV Report 2013, p. 256).

In 2013, 61 Nestlé factories, which accounts for 12 % of total factories, achieved zero waste for disposal. The objective we set in 2012 was achieved two years early (Nestlé CSV Report 2013, p. 207).

Between 2008 and 2013, the ratio of waste for disposal against raw material we used for production was reduced from 1.92 to 1.08 %. Table 7 presents the indicators regarding waste for disposal at Nestlé since 2008.

In comparison with the waste in the food manufacturing process, the amount of food waste generated in the early stages of the supply chain and consumers' consumption process is shocking. According to the Food and Agriculture Organization (FAO) report, each year about one-third of food produced for human consumption is lost or wasted globally. In medium and high-income countries, consumption stage is where food is wasted to a significant extent, while in low-income countries most of the food losses occur at the early stages of the food supply chain (Gustavsson et al. 2011).

Table 7 Indicators of material use efficiency (Nestlé CSV Report 2011, 2012, 2013)

Indicator	2008	2009	2010	2011	2012	2013
Waste for disposal (1000 tonnes)	410	359	370	343	315	257
Waste for disposal (kg/tonne product)	10	8.7	8.45	7.59	6.6	4.9
Hazardous waste (1000 tonnes)	n.a.	n.a.	n.a.	5.5	4.1	3.2
Waste and by-product recovery rate (%)	73	79	79	81	82	85
Number of zero waste factories	n.a.	n.a.	n.a.	n.a.	39	61
Raw material used (million tonnes excluding material for packaging purpose)	21.4	21.2	23.27	22.9	22.5	23.9
Waste ratio in production (%)	1.92	1.69	1.59	1.50	1.40	1.08

This issue is complex and involves various stakeholders at different stages. Nestlé carefully examined this issue and has adopted a holistic and collaborative approach to tackle this. At the early stage of the supply chain, Nestlé helps farmers reduce waste and loss during food harvesting and transportation processes. For example, in developing countries, we have provided cooling facilities to dairy farmers. These facilities have helped farmers cut milk lost between the farm and factory stage so huge amounts of water, energy and GHG emissions and money have been saved. In the district of Renala in Pakistan, more than half of milk losses have been saved at the early stage of the milk supply chain with the help of Nestlé (Nestlé CSV Report 2013, p. 258).

In Central and West Africa, contamination due to the humid environment conditions and poor drying and storage practices caused up to 30 % of cereal crop losses. The 'Grains Quality Improvement Project' was launched by Nestlé to help farmers in Ghana and Nigeria reduce mycotoxin contamination, in which 60,000 African farmers were trained in 2012 to produce grains with mycotoxin levels within Nestlé standards. The combination of toxin-reduction strategies such as good agricultural and storage practices, and capacity-building training sessions by Nestlé agronomists achieved a 60 % reduction in crop losses. (Nestlé CSV Report 2013, p. 259).

Downstream in the food value chain, Nestlé helps consumers reduce food wastage in the consumption process. For example, we have been developing creative solutions to help consumers make the most of leftovers. These include a range of different dough (pizzas, pasties, etc.) that can be filled with leftover food from their fridge, *Maggi* in France has brought out a smartphone app full of top tips and recipes on how to use leftovers (Nestlé CSV Report 2013, p. 260).

3.6 Resource Efficiency and Total Resource Consumption

To continue its efforts, Nestlé has achieved considerable success to improve its resource efficiency over the last decade. As shown in the Fig. 3, since 2003 Nestlé has reduced waste for disposal per tonne of product by 53.6 %, water withdrawal per tonne of product by 49.5 %, water discharge per tonne of product by 60.1 %, direct GHG emissions per tonne of product by 45.4 %, and total on-site energy consumption per tonne of product by 33.7 %. We have also reduced material use for packaging purpose per tonne of product by 0.2 % since 2009. Figure 3 shows Nestlé's resource use efficiency between 2003 and 2013 (data for material used for packaging is only available from 2009).

Due to our success in improving resource use efficiency, during the past ten years we have steadily decreased the total value of a few resources while increasing our production by 56 % in the same period. Our total water withdrawal has decreased by 21.2 %, water discharge reduced by 37.2 %, waste for disposal cut by 27.6 %, and direct GHG emissions fell by 14.9 %. We have kept our energy consumption almost unchanged in comparison to 2003. Figure 4 shows Nestlé's

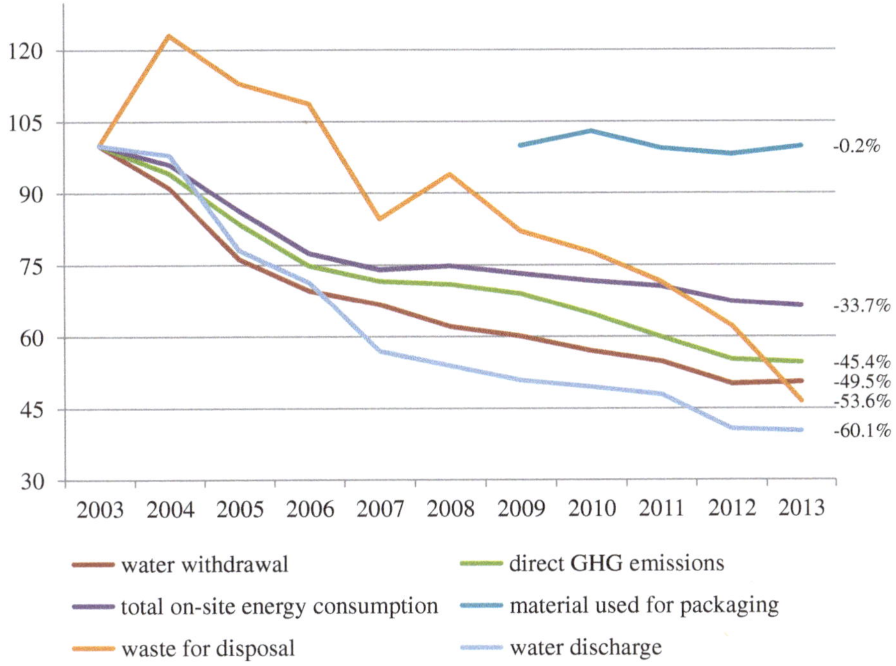

Fig. 3 Resource consumption and waste disposal per tonne product 2003–2013

total resource consumption between 2003 and 2013 (data for material used for packaging is available only from 2009).

Besides what is highlighted in Fig. 4, biodiversity, deforestation and lifecycle analysis are also important aspects to consider helping improve resource efficiency. More information on the efforts that Nestlé has made on these aspects can be found in Nestlé's CSV reports and corporate websites.

4 Conclusion

Within our core strategy framework of CSV, Nestlé has put tremendous and continuous efforts on improving resource efficiency over the last decade and steadily decreased its resource use, direct GHG emissions and waste for disposal, along with a steady increase in its manufacturing production.

This has resulted in a significant contribution to environmental sustainable development, and helping to protect the local and global environment. By improving efficiency and reducing waste and resource consumption, we have lowered the costs in our production to strengthen Nestlé's products competitiveness in the market.

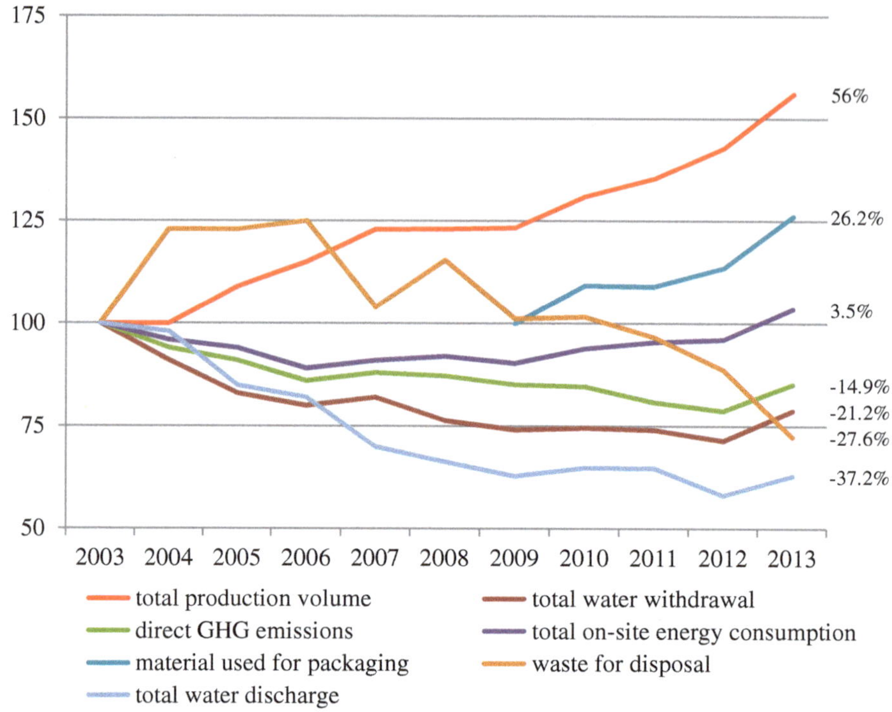

Fig. 4 Resource consumption and waste disposal vs production volume 2003–2013

Within the CSV framework, in the future we will continue to implement our strategies on environmental sustainability with an emphasis on energy efficiency at our factories, increase our use of cleaner fuels, invest in renewable energy sources such as solar and wind energy, improve material and water utilization efficiencies, reduce total waste for disposal, and increase the number of factories with zero waste for disposal. We will aim to do more to help farmers and consumers save resources both upstream and downstream in the value chain of our products. We will also continue to actively contribute to national and global partnerships to eliminate edible food wastage by addressing the perspectives of all stakeholders in the food chain.

While traditional resource-intensive growth is the ultimate root cause for the world's environmental problems, products with better environmental performance can be a significant growth driver for Nestlé's business. We strongly believe that our CSV strategy will bring a bright perspective to Nestlé's future development, benefiting both our shareholders and global sustainable development.

References

Abdullah MA, Agalgaonkar AP, Muttaqi KM (2014) Climate change mitigation with integration of renewable energy resources in the electricity grid of New South Wales, Australia. Renew Energy 66:305–313

Alcott B (2005) Jevons' paradox. Ecol Econ 54(1):9–21

Allen CD, Macalady AK, Chenchouni H, Bachelet D, McDowell N, Vennetier M et al (2010). A global overview of drought and heat-induced tree mortality reveals emerging climate change risks for forests. Forest Ecol Manage 259(4):660–684

Barrett J, Scott K (2012) Link between climate change mitigation and resource efficiency: a UK case study. Glob Environ Change 22(1):299–307

Brabeck-Letmathe P (2006) Interview with Kramer, M. [Online] Available from http:/// sharedvalue.org/sites/default/files/resource-files/Nestle_Corporate_Social_Responsibility_in_Latin_America.pdf

Coca-Cola—2011 Annual Report on Form 10-K (2011) [Online] Available from http://assets.coca-colacompany.com/b6/f3/ecad445f4fc1819dd37e04e057ad/form_10K_2011.pdf

Coca-Cola—2013 Annual Report on Form 10-K (2013) [Online] Available from http://assets.coca-colacompany.com/d0/c1/7afc6e6949c8adf1168a3328b2ad/2013-annual-report-on-form-10-k.pdf

Coca Cola - 2013/2014 Sustainable Report (2013) [Online] Available from: http://assets.coca-colacompany.com/1a/e5/20840408404b9bc484ebc58d536c/2013-2014-coca-cola-sustainability-report-pdf.pdf

Danone Annual Financial Report 2013 (2013) [Online] Available from http://www.danone.com/en/publications/

Danone Sustainability Report 2010 (2010) [Online] Available from http://www.danone.com/en/publications/

Danone Sustainability Report 2012 (2012) [Online] Available from http://www.danone.com/en/publications/

Danone Sustainability Report 2013 (2013) [Online] Available from http://www.danone.com/en/publications/

Fischer G, Tubiello FN, Van Velthuizen H, Wiberg DA (2007) Climate change impacts on irrigation water requirements: effects of mitigation, 1990–2080. Technol Forecast Soc Chang 74(7):1083–1107

General Mills Global Responsibility 2014 (2014) [Online] Available from http://www.generalmills.com/Company/publications

General Mills Annual Report 2013 (2013) [Online] Available from http://www.generalmills.com/Company/publications

Gustavsson J, Cederberg C, Sonesson U (2011) Global food losses and food waste [Online] Available from http://www.fao.org/fileadmin/user_upload/suistainability/pdf/Global_Food_Losses_and_Food_Waste.pdf

Jevons WS (1957) The theory of political economy, 5th edn. Augustus M. Kelley, New York

Kramer M (2006) [Online] Available from http://sharedvalue.org/sites/default/files/resource-files/Nestle_Corporate_Social_Responsibility_in_Latin_America.pdf

McMichael AJ, Woodruff RE, Hales S (2006) Climate change and human health: present and future risks. Lancet 367(9513):859–869

Nestec Ltd. (2013) The Nestlé Policy on Environmental Sustainability. [Online] Available from http://www.nestle.com/asset-library/documents/library/documents/environmental_sustainability/nestl%C3%A9%20policy%20on%20environmental%20sustainability.pdf

Nespresso. Designing more efficient machines. [Online] Available from http://www.nestle-nespresso.com/ecolaboration/case-studies/designing-more-efficient-machines

Nespresso. An international joint effort to create shared value in Jardín. [Online] Available from http://www.nestle-nespresso.com/sustainability/case-studies/community-milling-in-colombia

Nestlé Annual Report 2011 (2011) [Online] Available from http://www.nestle.com/asset-library/Documents/Library/Documents/Annual_Reports/2011-Annual-Report-EN.pdf
Nestlé Annual Report 2013 (2013) [Online] Available from http://www.nestle.com/asset-library/documents/library/documents/annual_reports/2013-annual-report-en.pdf
Nestlé Creating Shared Value Report 2011 (2011) [Online] Available from http://www.nestle.com/asset-library/Documents/Library/Documents/Corporate_Social_Responsibility/2011-CSV-report.pdf
Nestlé Creating Shared Value Report 2012 (2012) [Online] Available from http://www.nestle.com/asset-library/documents/library/documents/corporate_social_responsibility/nestle-csv-full-report-2012-en.pdf
Nestlé Creating Shared Value Report 2013 (2013) [Online] Available from: http://www.nestle.com/asset-library/documents/library/documents/corporate_social_responsibility/nestle-csv-summary-report-2013-en.pdf
Nestlé UK. *Achieving Zero Waste to Landfill.* [Online] Available from http://www.nestle.com/csv/case-studies/AllCaseStudies/Nestle-UK-Achieving-Zero-Waste-to-Landfill
Project News Report: Nestlé's Sustainable Agriculture Initiatives [Online] Available from http://thenest-eur-hq.nestle.com/TP/TP_AGRI/Documents%20AGR/06%20Sustainable%20Agriculture/SAIN%20Cases/2011/005%202011%20SAIN%20Water%20Colombia%20Nespresso%20Community%20Mill.pdf
Nestlé Waters Creating Shared Value Report 2011 (2011) [Online] Available from http://www.nestle-waters.com/content/documents/css/images/pdf/nestl%C3%A9_waters_csv_report_2011.pdf
Olesen JE, Bindi M (2002) Consequences of climate change for European agricultural productivity, land use and policy. Eur J Agron 16(4):239–262
PepsiCo Sustainability Report 2011/2012 (2012) [Online] Available from http://www.pepsico.com/Assets/Download/PEP_CSR12_2011-2012_Sustainability_Report.pdf
PepsiCo Sustainability Report 2013 (2013) [Online] Available from http://www.pepsico.com/Assets/Download/PEP_2013_Sustainability_Report.pdf
PepsiCo 2011 Annual Report (2011) [Online] Available from http://www.pepsico.com/annual11/downloads/PEP_AR11_2011_Annual_Report.pdf
PepsiCo 2013 Annual Report (2013) [Online] Available from http://www.pepsico.com/Assets/Download/PEP_Annual_Report_2013.pdf
Porter ME (2011) [Online] Available from http://www.nespresso.com/ecolaboration//medias_dyn/articles/1791/article/attachment-2.pdf
Porter ME, Kramer MR (2011) The big idea: Creating shared value. Harvard Bus Rev 89(1):2
Tyson—2012 Annual Report on Form 10-K (2012) [Online] Available from http://ir.tyson.com/files/doc_financials/2012/Tyson-form10k_092912.pdf
Tyson—2013 Annual Report on Form 10-K (2013) [Online] Available from http://ir.tyson.com/files/doc_financials/2013/TSN-2013-Annual-Report-on-Form-10-K.PDF
Tyson Foods Sustainability Report 2013 (2013) [Online] Available from http://www.interfacecutthefluff.com/wp-content/uploads/2012/09/2013-Sustainability-Report-Executive-Summary.pdf
United States Millennium Development Goals Database [Online] Available from http://data.un.org/Data.aspx?q=carbon+dioxide&d=MDG&f=seriesRowID%3a776

Water Resources Management and Adaptation to Climate Change

Taikan Oki

Abstract Initiatives to solve world water issues and climate change have been individually developed and implemented without adequate synergy among them. The relevance of adapting to climate change has increased in addition to the mitigation to climate change. Integrated water resources management (IWRM) is expected to work well as an adaptation to climate change, preferably integrated with disaster risk management (DRM), proper land management, and poverty alleviation, to accomplish sustainable development. A synthesized review on water and climate is provided based on the 5th Assessment of the Intergovernmental Panel on Climate Change.

1 Introduction

Climate change due to the increase of greenhouse gases (GHGs) in the atmosphere has been a concern among scientists even since the late 19th century (Tyndall 1861; Arrhenius 1896; Callendar 1938; Manabe and Strickler 1964). Initiated by the International Geophysical Year (IGY) in 1958, measurements of the concentration of carbon dioxide (CO_2) in the atmosphere started at the observatories in Antarctica and the top of Mauna Loa Mountain in Hawaii (Pales and Keeling 1965), and a significant increase of the CO_2 concentration, with predominant seasonal variation due to vegetation uptake and decomposition, was affirmed. The latest news is that global mean concentration of CO_2 in the atmosphere is estimated to be over 400 ppm ever since these observations started.

However, climate change was not the major concern in the international community both politically and economically until the end of the Cold War around 1990. Then, climate change became one of the major issues in the United Nations Conference on Environment and Development, sometimes called the "Earth Summit", held in Rio de Janeiro, Brazil, in 1992. The international community

T. Oki (✉)
Institute of Industrial Science, The University of Tokyo,
4-6-1 Komaba, Meguro-Ku, Tokyo 153-8505, Japan
e-mail: taikan@iis.u-tokyo.ac.jp

A.K. Biswas and C. Tortajada (eds.), *Water Security, Climate Change and Sustainable Development*, Water Resources Development and Management, DOI 10.1007/978-981-287-976-9_3

recognized climate change as one of the major obstacles for sustainable development of the world, and climate change became the most challenging issue to be tackled worldwide.

Water was also one of the global issues included in the Agenda 21, the action plan adopted in the "Earth Summit". However, world water issues and climate change were not synergistically discussed during the 20th century, even though water and hydrological cycles are the major component of the climate systems (Oki et al. 2004).

At the beginning of the 21st century, assessments of the world water resources considering projected climate changes started (Vörösmarty et al. 2000, 2010; Oki et al. 2001; Arnell 2004; Oki and Kanae 2006; Alcamo et al. 2007; Hanasaki et al. 2013a, b), with populaization of the concept of "Stationarity is dead" (Milly et al. 2008), suggesting water resources management should be fundamentally reevaluated because natural systems do not fluctuate within an unchanging envelope of variability observed in the past.

The 5th assessment reports (AR5) of the Intergovernmental Panel on Climate Change have been published by Working Group I (IPCC 2013), WGII (IPCC 2014a), and WGIII (IPCC 2014b), and the Synthesis Report (SYR) was also published in 2014 (IPCC 2014c). In this review article, the latest scientific basis on the climate change, its impacts, and the adaptation, particularly related to freshwater resources and their managements, are introduced mainly based on IPCC AR5. Another review of AR5/WGII Chap. 3 "Freshwater Resources" can be found in Döll et al. (2015).

2 Observed and Projected Climate Changes

Two and a half decades ago, human beings were still uncertain about the anthropogenic changes of the climate system even in terms of the rise in global mean temperature. The First Assessment Report (FAR) of IPCC WGI published in 1990 starts from a sentence saying "We are certain of the following: …emissions resulting from human activities are substantially increasing the atmospheric concentrations of the greenhouse gases. These increases will enhance the greenhouse effect, resulting on average in an additional warming of the Earth's surface", but pessimistically says "The unequivocal detection of the enhanced greenhouse effect from observations is not likely for a decade or more" (IPCC 1990). Unfortunate for the natural and human systems, the temperature rise especially in the 1990s was intensive, and the latter reports of the IPCC could convincingly detect the enhanced greenhouse effect on the global mean temperature.

Five years later, however, the Second Assessment Report (SAR) of IPCC still modestly stated "The balance of evidence suggests a discernible human influence on global climate" (IPCC 1995), and the Third Assessment Report (TAR) provided a little stronger message in 2001, saying "There is new and stronger evidence that most of the warming observed over the last 50 years is attributable to human activities" (IPCC 2001). In 2007, the Fourth Assessment Report (AR4) articulated as "Most of the observed increase in global average temperatures since the mid-20th century is very likely due to the observed increase in anthropogenic greenhouse gas concentrations" (IPCC 2007), and the Fifth Assessment Report (AR5) of WGI (IPCC 2013) supported the message by sentences "warming of the climate system is unequivocal" and "It is extremely likely that human influence has been the dominant cause of the observed warming since the mid-20th century".

This increment of confidence on the changing climate and its attribution to anthropogenic impacts is the result of observational evidences and the progress in climate modelling during the last three decades. However, compared to temperature, natural variation is larger and the long-term observational record is not sufficient for precipitation, particularly during the first half of the 20th century outside of developed countries. Consequently, the observed changes of precipitation are described with lower confidence even in AR5/WGI as "Confidence in precipitation change averaged over global land areas since 1901 is low prior to 1951 and medium afterwards" (IPCC 2013). Despite this, increasing trends are detected in the mean precipitation averaged over the mid-latitude land areas of the Northern Hemisphere, while this remains uncertain for other latitudes.

Further, uncertainties in extreme precipitation are larger and assessed as "There are likely more land regions where the number of heavy precipitation events has increased than where it has decreased. The frequency or intensity of heavy precipitation events has likely increased in North America and Europe" (IPCC 2013). This phrase should not be understood that extreme precipitation in North America and Europe has increased more than in other continents. This is because the detection and attribution of the changes in extreme precipitation in other continents were impossible, due to the shortage of historical records.

Nevertheless, AR5/WGII delivers a message:

> In many regions, changing precipitation or melting snow and ice are altering hydrological systems, affecting water resources in terms of quantity and quality (medium confidence). Glaciers continue to shrink almost worldwide due to climate change (high confidence), affecting runoff and water resources downstream (medium confidence). Climate change is causing permafrost warming and thawing in high latitude regions and in high-elevation regions (IPCC 2014a).

The difference of the tones between WGI and WGII on the attribution of the changes in hydrological cycles to climate change is partially due to the difference in their basic philosophies that WGI quantifies the links between observed climate change and human activity, as well as other external climate drivers. On the contrary, WGII considers the links between impacts on natural and human systems and observed climate change, regardless of its cause and attribution to anthropogenic activities.

3 Projected Climate Changes

Changes in hydrological cycles associated with the anthropogenic climate change are projected as:

> Changes in the global water cycle in response to the warming over the 21st century will not be uniform. The contrast in precipitation between wet and dry regions and between wet and dry seasons will increase, although there may be regional exceptions (IPCC 2013).

Since quantitative "water resources development" of surface water is equivalent to regulating and levelling the spatial heterogeneity and temporal variation of available water resources by long-range transport and/or storage and release, intensification of the contrast of wet and dry regions/seasons will exacerbate the delicate balance between water demand and supply at present.

In terms of the projection of extreme precipitation, AR5/WGI concludes "Extreme precipitation events over most of the mid-latitude land masses and over wet tropical regions will very likely become more intense and more frequent by the end of this century, as global mean surface temperature increases" (IPCC 2013). The fundamental mechanism driving the changes in precipitation behaviour is the changes in the moisture content of the atmosphere governed by the Clausius–Clapeyron relation, with scaling of the precipitation intensity of approximately 7 % per degree Celsius. As Utsumi et al. (2011) unveiled, intensity of short-term (e.g., 10 min) precipitation observed in the past increased following the Clausius-Clapeyron relation, and continuous temperature rise due to anthropogenic climate change will have a potential to increase the risk of urban flush floods which can be triggered by heavy rainfall in very short term.

More frequent precipitation with certain intensity and more intense precipitation with certain probability of exceedance are the two sides of a coin as illustrated in Fig. 1. Probability distribution functions were fit for annual maximum daily precipitation near Tokyo, simulated by a general circulation model (GCM; Numaguti et al. 1997), and intensities of daily precipitation for each return period are plotted for the 20th century and the 21st century simulations. Daily precipitation intensity with once in 100 year probability of exceedance (annually 1 %) during the 20th century was estimated as 77.7 mm day^{-1}, and it is projected to be 84.1 mm day^{-1} in the 21st century. The increase of the intensity is approximately 10 %, however, 84.1 mm day^{-1} in the 20th century corresponds approximately to once in 300 years (annually 0.3 %). In other words, the frequency increases approximately 3 times in this case, and 77.7 mm day^{-1}, once in 100 year during the 20th century, corresponds approximately to once in 30 years in the 21st century. A more comprehensive analysis based on multiple GCM simulations is presented in the Special Report of IPCC "Managing the Risks of Extreme Events and Disasters to Advance Climate Change Adaptation" (SREX; IPCC 2012).

As for monsoon systems, "Globally, it is likely that the area encompassed by monsoon systems will increase over the 21st century." (IPCC 2013), but the monsoonal wind will be weakened due to the decrease of the temperature difference between continents and oceans. Onset dates of monsoon systems are projected to be earlier or not to change much, retreat dates are projected to be delayed, and consequently the length of the monsoon season will be longer in many regions. Monsoonal precipitation is projected to be intensified due to the increase of water vapor in the atmosphere.

Tropical cyclones have potentials to trigger serious disaster such as inland and coastal flooding or storm surge, but at the same time, the associated rainfall with

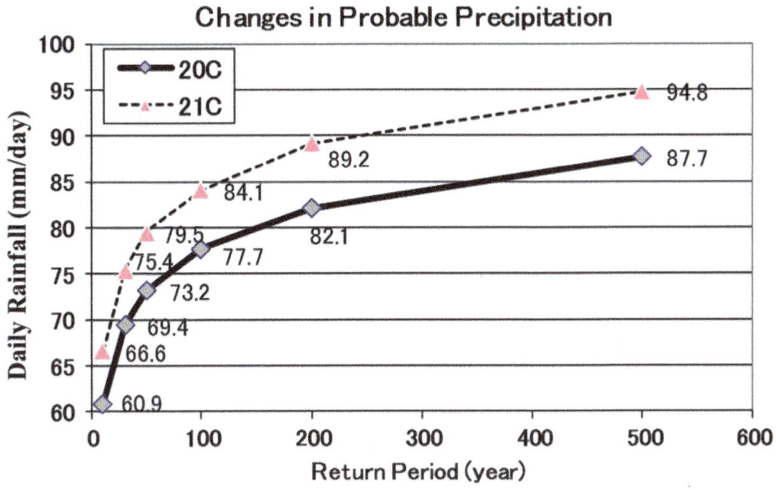

Fig. 1 Changes in precipitation rate (ordinate axis) corresponding to return period (abscissa axis) during the 20th century (*thick line* with *rectangles*) and the 21st century (*thin dotted line* with *triangles*) estimated by fitting probability distribution functions for annual maximum daily rainfall near Tokyo simulated by a general circulation model

tropical cyclones can be a major source of water resources in some regions. AR5/WGI says "it is likely that the global frequency of occurrence of tropical cyclones will either decrease or remain essentially unchanged, concurrent with a likely increase in both global mean tropical cyclone maximum wind speed and precipitation rates" (IPCC 2013), namely negative impacts, causing damages due to strong winds, torrential rainfall, and higher storm surges, would outweigh the positive impacts to increase available water resources by tropical cyclones.

4 Key Regional Risks and Potentials for Reducing Risks

It should be noted that increasing frequency of natural hazards, such as extreme rainfall or long-lasting heat waves, alone will not cause damage on human and natural systems, and both climate and social changes are relevant for managing global risks and developing sustainability for the future, as articulated in the AR5 of IPCC WGII, "Risk of climate-related impacts results from the interaction of climate-related hazards (including hazardous events and trends) with the vulnerability and exposure of human and natural systems" (IPCC 2014a).

Corresponding to the projected climate changes and associated impacts, a few key risks and potential adaptation measures to reduce the risks are identified mainly by expert judgements and presented in SPM 2 Table of AR5/WGII (IPCC 2014a). The concept of risk and risk management became the main sheme in IPCC reports since SREX (IPCC 2012), which defines (disaster) risk as "emerges from the interaction of weather or climate events, the physical contributors to disaster risk, with exposure and vulnerability, the contributors to risk from the human side." In AR5/WGII, risk is defined as "The potential for consequences where something of value is at stake and where the outcome is uncertain, recognizing the diversity of values." Therefore "impacts" caused by climate changes or "sectors" vulnerable for climate changes are mainly adopted in SPM2 Table of AR5/WGII (IPCC 2014a), instead of "risks" which is referred as "often represented as probability of occurrence of hazardous events or trends multiplied by the impacts if these events or trends occur" (IPCC 2014a).

- Africa: water resources, food security, and vector- and water-borne disease
- Europe: flooding, water restrictions, and extreme heat events
- Asia: flooding, heat-related mortality, and drought-related water and food shortage
- Australasia: coral reef systems, flood damage, and coastal infrastructure and low-lying ecosystems
- North America: wild fire, heat-related human mortality, and urban floods
- Central and South America: water availability, flooding and landslides, food, and vector-borne diseases
- Polar Regions: freshwater and terrestrial ecosystems, health and well-being, and northern communities
- Small islands: livelihoods and coastal areas
- Ocean: fisheries catch potential, coral reefs, and sea level rise

It should be noted that water is related to most of the key risks in each region, suggesting that water is the delivering mechanism of climate change impacts to society. In terms of the costs of adaptation, the water sector requires more costs than agriculture/forestry/fisheries and human health sectors, but infrastructure and coastal zones would require more in developing countries (Chambwera et al. 2014).

Some adaptation measures are region or sector specific. However, in general, technological measures, such as flood protection and urban drainage, could be effective but with high costs. Improved or new technological developments towards new crop varieties more adapted to climate change and early warning systems are also expected as adaptation. Institutional measures, such as land use planning, appropriate building codes, and integrated water resources management, should also be considered as options of adaptation to climate change. Reducing other stressors would be an alternative choice as an adaptation measure where the potential of direct actions to reduce the anticipated impacts due to climate change is limited.

5 Impacts of Climate Change on Freshwater Resources

A synthesis report (IPCC 2014c) delivered the headline "Continued emission of greenhouse gases will cause further warming and long-lasting changes in all components of the climate system, increasing the likelihood of severe, pervasive and irreversible impacts for people and ecosystems" and this statement is supported by the finding that "Global mean temperature is projected to rise approximately proportional to the cumulative total anthropogenic CO_2 emissions from 1870" (IPCC 2013).

Temperature rise itself will have direct impacts on the availability of water resources through changing flow regimes in snow-dominant or glacier-effluent river basins. Ironically, in these rivers, "total meltwater yields from stored glacier ice will increase in many regions during the next decades but decrease thereafter" (Jiménez Cisneros et al. 2014). Higher air temperature will increase the temperature of surface water, and it will have negative impacts on freshwater ecosystems and "reduce raw water quality, posing risks to drinking water quality even with conventional treatment" (Jiménez Cisneros et al. 2014).

Sea level rise, which is predominantly driven by temperature rise through ocean thermal expansion and glacier mass loss, will accelerate sea water intrusions to surface and ground water and reduce freshwater availability. Increase of mean sea level will be associated with the increase of extremely high sea level, and increase the risk of coastal flooding and storm surge. Further, climate change is projected to alter hydrological cycles: changing temporal and geographical patterns of hydrological components and their extremes.

Consequently, "Freshwater-related risks of climate change increase significantly with increasing greenhouse gas (GHG) concentrations" (Jiménez Cisneros et al. 2014), and the frequency of floods and/or droughts is projected to increase in some parts of the world. There are only few studies quantifying the relationship between GHG concentration pathways and water-related damage. Hirabayashi et al. (2013) estimated that the fraction of the world population exposed to a 20th century 100-year flood is three times higher for RCP8.5 than for RCP2.6 at the end of the 21st century.

Particularly, in most dry subtropical regions, "Climate change is projected to reduce renewable surface water and groundwater resources significantly" (Jiménez Cisneros et al. 2014), and it is concerned to "intensify competition for water among agriculture, ecosystems, settlements, industry, and energy production, affecting regional water, energy, and food security" (Jiménez Cisneros et al. 2014). This is mainly because of the intensification of the temporal variability of precipitation, and also the increase of the evaporation loss from reservoirs.

Climate change will also have adverse impacts on the demand side of water resources. Irrigation water demand will increase in many areas due to the projected air temperature rise and the changes in precipitation, even though the increase would be moderated by the elevated CO_2 concentration in the atmosphere since the transpiration efficiency associated with photosynthesis will be improved. Higher temperature increases water demand for municipal services, as well.

The energy sector will also be affected by changes in the hydrological cycles associated with climate change in terms of the availability of water resources for cooling of thermal power plants, and hydropower generation.

6 Adaptation for Climate Change in Water Sectors

Mitigation for climate change, reducing the emission of GHGs to slow down the speed of climate change, has been at the centre of the actions of tackling the projected climate change impacts. However, since AR4 (IPCC 2007), it has been recognized that there are definite climate changes due to the already emitted GHGs and slow responses of the climate systems, mainly oceanic and glacier processes; climate change will persist for many centuries even if emissions of GHGs are stopped now (IPCC 2013).

As a consequence, adaptation to climate change impacts became one of the complementary strategies for reducing and managing risks of climate change, with conditional expressions, such as "Greater rates and magnitude of climate change increase the likelihood of exceeding adaptation limits (limits to adaptation)" and "Many adaptation and mitigation options can help address climate change, but no single option is sufficient by itself" (IPCC 2014c).

Basic concept of adaptation is to reduce the vulnerabilities of natural and built environments and the exposure of ecosystems and human settlements for climate change related hazards, and minimize the risk. Detailed options of adaptation for climate change relevant for freshwater resources management are given in Table 3-3 of AR5/WGII (IPCC 2014a), and some examples are introduced here.

First of all, institutional measures of adaptation for climate change are implementing integrated water resources management, building of adaptive capacity, development of financial tools, such as credit, subsidies, and public investment, and promoting synergy of water and energy savings and efficient use.

In the field of design and operation, revising the design criteria of water infrastructure to optimize flexibility, redundancy, and robustness, diversifying water sources and improving reservoir management, reducing demand by controlling leaks, implementing water-saving programs, and cascading and reusing water would be adaptation measures in the water sectors.

To implement monitoring and early warning systems, to design cities and rural settlements more resilient to floods, to promote switching to more appropriate crops, such as drought-resistant, salt-resistant, and low water demand, and to plant flood- or drought-resistant crop varieties would reduce impacts of natural disasters and are adaptations to climate change.

In the agricultural sector, improvement of irrigation efficiency, reduction of demand for irrigation water, and reuse of wastewater to irrigate crops are possible adaptation measures.

In the industrial sector, relocation of water-thirsty industries and crops to water-rich areas, and implementation of industrial water efficiency certifications can be adaptations to climate change.

As can be seen from the examples, "a first step towards adaptation to future climate change is reducing vulnerability and exposure to present climate variability" (IPCC 2014a). As impacts from recent climate-related extremes, such as heat waves, droughts, floods, cyclones, and wildfires, reveal significant vulnerability and

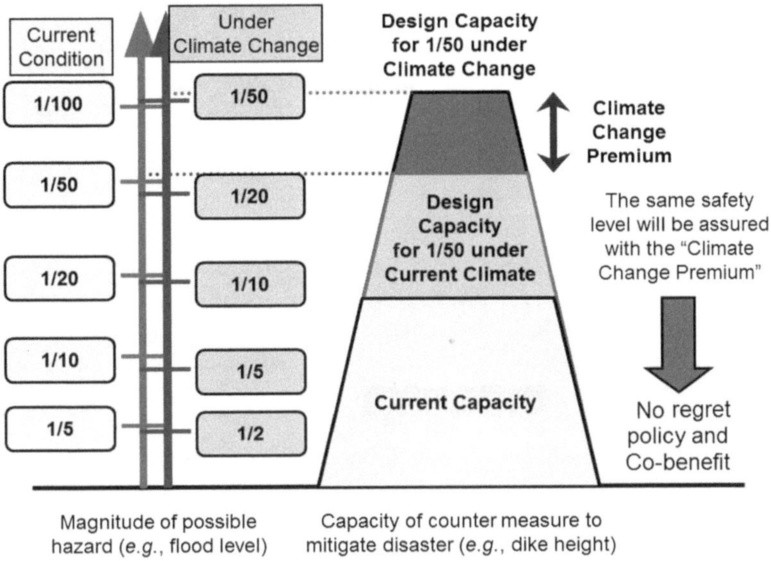

Fig. 2 Illustration of the impacts of climate change on the design level in water resources management. Safety level will be reduced when the same magnitude of hydro-meteorological hazard will occur more frequently due to changing climate, and the same level of safety will be assured by some increased adaptation capacity, referred to as "climate change premium". In the regions vulnerable for current climate variability, increasing the safety level has low potential to cause regrets in the future

exposure of some ecosystems and many human systems to current climate variability, current resilience and preparedness for present climate variations are generally not satisfactory in most parts of the world. Therefore, reducing vulnerability and exposure would be beneficial anyway and they are in line with the low-regret policy (Fig. 2). To avoid maladaptation, or adaptation which goes wrong, "scientific research results should be analyzed during planning" (IPCC 2014a).

7 Practical Adaptation Measures in Water Sectors in Developing States Such as India

Adaptation measures to climate change in the water sector are basically alike ordinary water resource development, flood disaster mitigation, and conservation of aquatic ecosystems with different motivations caused by anticipated climatic changes and with huge uncertainties in the changes. It should be noted that the huge uncertainties in climate change projections should not be used as an excuse to suspend decision making and not to take actions, because most of the decision making now is also based on uncertain facts and insufficient evidence.

Ground water reserves are reliable sources of water supply if managed and exploited in a sustainable way. Compared to surface water, ground water has a larger capacity to regulate temporal variability of hydrological cycles, and should be less susceptible to variability of hydrological cycles due to anticipated climate changes. Rainwater harvesting by facilities promoting recharge into aquifers, would enhance the tolerance of ground water usage against climatic variation, and would be a reasonable adaptation measure in developing countries. Small-scale ponds, even tanks and lakes would work in a similar way.

Large-scale reservoirs would of course be able to regulate climatic variabilities if the cost is feasible and associated impacts are acceptable. Water transfer by long-range channels or even by other means such as a big plastic bag towed by a boat would smooth the spatial heterogeneity of freshwater resources, increase the availabilities of them, and could be one of the possible adaptation measures if being feasible economically and environmentally.

The cost of desalination has been declining and desalination of sea water would be feasible as for an adaptation measure in water-scarce regions even in developing countries. However, desalination demands substantial energy and would have a trade-off against mitigation of climate change. Compared to the desalination of sea water, recycling of used water, such as the reuse of purified sewage water for irrigation or inflow into reservoirs or aquifers, has lower cost and lower energy demand.

Demand side controls, such as adequate pricing for water and water usage, economic incentives to reduce water consumptions, and improving water-use efficiencies are also possible adaptation measures for climate change in water sectors. Importing water-intensive goods, typically food, can reduce local demand for water resources to produce the goods. From this point of view, food trade is sometimes called virtual water trade and can be considered as an adaptation measure in the water resources sector (Allan 1998; Oki and Kanae 2004).

For flood risk management, developing early warning systems with preparing shelters and better land use planning and regulations to prohibit from utilizing flood-prone lands for residential and industrialized areas would efficiently act as adaptation measure without intensive construction of infrastructure such as embankments and reservoirs.

8 Concluding Remarks

"People and societies may perceive or rank risks and potential benefits differently, given diverse values and goals" as stated in the 2nd paragraph of the AR5/WGII SPM (IPCC 2014a), and "climate-related hazards exacerbate other stressors, often with negative outcomes for livelihoods, especially for people living in poverty" (IPCC 2014a). These messages are rephrased as "climate change will amplify existing risks and create new risks for natural and human systems. Risks are unevenly distributed and are generally greater for disadvantaged people and communities in countries at all levels of development" in the AR5/SYR (IPCC 2014c). AR5/WGII SPM (IPCC 2014a) also states "For most economic sectors, the impacts of drivers such as changes in population, age structure, income, technology, relative prices, lifestyle, regulation, and governance are projected to be large relative to the impacts of climate change". Therefore, tackling climate change would not necessarily be a top priority issue for a significant number of people in the world, and it should not be an easy task to develop a consensus on the appropriate balance between mitigation and adaptation to climate change, even though tackling other issues than climate change would also have synergetic benefit for adaptation to climate change.

AR5/WGII pointed out that "significant co-benefits, synergies, and tradeoffs exist between mitigation and adaptation and among different adaptation responses; interactions occur both within and across regions" (IPCC 2014a), and mitigation and/or adaptation actions should not be planned in an isolated manner. Moreover, adaptation to climate change should be integrated into wider frameworks, such as disaster risk management, integrated water resources management, and sustainable development considering the limited human, institutional, and financial resources. All of them should preferably be integrated into a risk management framework assessing and managing possible global and local risks, and ultimately pursue the increase of human well-beings. Some pioneering examples are introduced in AR5/WGII (IPCC 2014a), and expressed as "adaptation is becoming embedded in some planning processes".

Food, energy, and water nexus (FEW Nexus) should also be noted in terms of the impacts of climate change, and mitigation and adaptation measures. There are tight connections, inter-dependencies, complementarities, trade-offs, and competitions among food, energy, and water uses, securities, and their changes under climate change (Fig. 3). The major common factor controlling the sustainability and security of the FEW Nexus is land, therefore land use/land cover management should also be integrated in the global risk management framework of FEW Nexus and climate change. A more detailed introduction can be found in the cross-chapter box "The Water–Energy–Food/Feed/Fibre Nexus as Linked to Climate Change" in AR5/WGII (Arent et al. 2014).

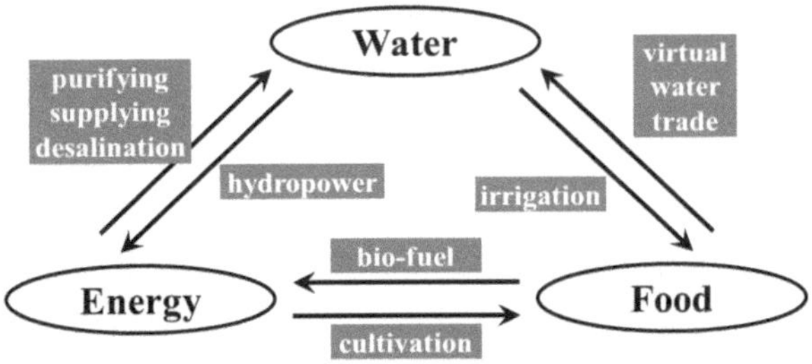

Fig. 3 Tight connections, inter-dependencies, complementarities, trade-offs, and competitions among food, energy, and water uses and their changes under climate change (FEW-Nexus)

Water cannot be secured by its own, and should be placed in a wider context, such as climate change adaptation, disaster risk management, and sustainable development. Further detailed and both quantitative and qualitative studies are expected to provide deeper insight on the issue.

Acknowledgements This work was supported by Grant-in-Aid for Scientific Research (S) (23226012) of JSPS, SOUSEI Programme of the Ministry of Education, Culture, Sports, Science, and Technology (MEXT), and the Environment Research and Technology Development Fund (S-10) of the Ministry of the Environment, Japan.

References

Alcamo J, Flörke M, Märker M (2007) Future long-term changes in global water resources driven by socio-economic and climatic changes. Hydrol Sci J 36(4):545–546

Allan JA (1998) Virtual water: a strategic resource. Global solution to regional deficits. Groundwater 36(4):545–546

Arent, DJ, Döll P, Strzepek KM, Jiménez Cisneros BE, Reisinger A, Tóth FL, Oki T (2014) Cross-chapter box on the water–energy–food/feed/fiber nexus as linked to climate change. In: Field CB, Barros VR, Dokken DJ, Mach KJ, Mastrandrea MD, Bilir TE, Chatterjee M, Ebi KL, Estrada YO, Genova RC, Girma B, Kissel ES, Levy AN, MacCracken S, Mastrandrea PR, White LL (eds) Climate change 2014: impacts, adaptation, and vulnerability. Part A: global and sectoral aspects. Contribution of working group II to the fifth assessment report of the intergovernmental panel on climate change. Cambridge University Press, Cambridge, pp. 163–166

Arnell NW (2004) Climate change and global water resources: SRES emissions and socio-economic scenarios. Global Environ Change 14:31–52

Arrhenius S (1896) On the influence of carbonic acid in the air upon the temperature of the ground. Philos Mag 41:237–276

Callendar GS (1938) The artificial production of CO_2 and its influence on temperature. Q J R Meteorol Soc 64:223–237

Chambwera M, Heal G, Dubeux C, Hallegatte S, Leclerc L, Markandya A, McCarl BA, Mechler R, Neumann JE (2014) Economics of adaptation. In: Field CB, Barros VR, Dokken DJ, Mach KJ, Mastrandrea MD, Bilir TE, Chatterjee M, Ebi KL, Estrada YO, Genova RC, Girma B, Kissel ES, Levy AN, MacCracken S, Mastrandrea PR, White LL (eds) Climate change 2014: impacts, adaptation, and vulnerability. Part A: global and sectoral aspects. Contribution of working group II to the fifth assessment report of the intergovernmental panel on climate change. Cambridge University Press, Cambridge, pp. 945–977

Döll P, Jimenez-Cisneros B, Oki T, Arnell NW, Benito G, Cogley JG, Jiang T, Kundzewicz ZW, Mwakalila S, Nishijima A (2015) Integrating risks of climate change into water management. Hydrol Sci J 60(1):4–13

Hanasaki N, Fujimori S, Yamamoto T, Yoshikawa S, Masaki Y, Hijioka Y, Kainuma M, Kanamori Y, Masui T, Takahashi K, Kanae S (2013a) A global water scarcity assessment under shared socio-economic pathways—part 1: water use. Hydrol Earth Syst Sci 17: 2375–2391

Hanasaki N, Fujimori S, Yamamoto T, Yoshikawa S, Masaki Y, Hijioka Y, Kainuma M, Kanamori Y, Masui T, Takahashi K, Kanae S (2013b) A global water scarcity assessment under shared socio-economic Pathways—part 2: water availability and scarcity. Hydrol Earth Syst Sci 17:2393–2413

Hirabayashi Y, Mahendran R, Koirala S, Konoshima L, Yamazaki D, Watanabe S, Kanae S (2013) Global flood risk under climate change. Nature Clim Change 3:816–821

IPCC (1990) Climate change: the IPCC scientific assessment. In Houghton JT, Jenkins GJ, Ephraums JJ (eds) Cambridge University Press, Cambridge, UK, 365 pp

IPCC (1995) Climate change 1995: the science of climate change. In: Houghton JT, Meira Filho LG, Callander BA, Harris N, Kattenberg A, Maskell K (eds) Cambridge University Press, Cambridge, UK, 572 pp

IPCC (2001) Climate change 2001: the scientific basis. In: Houghton JT, Ding Y, Griggs DJ, Noguer M, van der Linden PJ, Dai X, Maskell K, Johnson CA (eds) Contribution of working group I to the third assessment report of the intergovernmental panel on climate change. Cambridge University Press, Cambridge, 881 pp

IPCC (2007) Summary for policymakers. In: Solomon S, Qin D, Manning M, Chen Z, Marquis M, Averyt KB, Tignor M, Miller HL (eds) Climate change 2007: the physical science basis. Contribution of working group I to the fourth assessment report of the intergovernmental panel on climate change. Cambridge University Press, Cambridge

IPCC (2012) Summary for policymakers. In: Field CB, Barros VR, Dokken DJ, Mach KJ, Mastrandrea MD, Bilir TE, Chatterjee M, Ebi KL, Estrada YO, Genova RC, Girma B, Kissel ES, Levy AN, MacCracken S, Mastrandrea PR, White LL (eds) Managing the risks of extreme events and disasters to advance climate change adaptation. A special report of working groups I and II of the intergovernmental panel on climate change. Cambridge University Press, Cambridge, pp 1–19

IPCC (2013) Summary for policymakers. In: Stocker VTF, Qin D, Plattner G-K, Tignor M, Allen SK, Boschung J, Nauels A, Xia Y, Bex V, Midgley PM (eds) Climate change 2013: the physical science basis. Contribution of working group I to the fifth assessment report of the intergovernmental panel on climate change. Cambridge University Press, Cambridge

IPCC (2014a) Summary for policymakers. In: Field CB, Barros VR, Dokken DJ, Mach KJ, Mastrandrea MD, Bilir TE, Chatterjee M, Ebi KL, Estrada YO, Genova RC, Girma B, Kissel ES, Levy AN, MacCracken S, Mastrandrea PR, White LL (eds) Climate change 2014: impacts, adaptation, and vulnerability. Part A: global and sectoral aspects. Contribution of working group II to the fifth assessment report of the intergovernmental panel on climate change. Cambridge University Press, Cambridge, pp. 1–32

IPCC (2014b) Summary for policymakers. In: Edenhofer O, Pichs-Madruga R, Sokona Y, Farahani E, Kadner S, Seyboth K, Adler A, Baum I, Brunner S, Eickemeier P, Kriemann B, Savolainen J, Schlömer S, von Stechow C, Zwickel T, Minx JC (eds) Climate change 2014: mitigation of climate change. Contribution of working group III to the fifth assessment report of the intergovernmental panel on climate change. Cambridge University Press, Cambridge

IPCC (2014c) Climate change 2014: synthesis report. In: Core Writing Team, Pachauri RK, Meyer LA (eds) Contribution of working groups I, II and III to the fifth assessment report of the intergovernmental panel on climate change. IPCC, Geneva, Switzerland, 151 pp

Jiménez Cisneros BE, Oki T, Arnell NW, Benito G, Cogley JG, Döll P, Jiang T, Mwakalila SS (2014) Freshwater resources. In: Field CB, Barros VR, Dokken DJ, Mach KJ, Mastrandrea MD, Bilir TE, Chatterjee M, Ebi KL, Estrada YO, Genova RC, Girma B, Kissel ES, Levy AN, MacCracken S, Mastrandrea PR, White LL (eds) Climate change 2014: impacts, adaptation, and vulnerability. Part A: global and sectoral aspects. Contribution of working group II to the fifth assessment report of the intergovernmental panel on climate change. Cambridge University Press, Cambridge, pp 229–269

Manabe S, Strickler RF (1964) On the thermal equilibrium of the atmosphere with a convective adjustment. J Atmos Sci 21:361–385

Milly PCD, Betancourt J, Falkenmark M, Hirsch RM, Kundzewicz ZW, Lettenmaier DP, Stouffe RJ (2008) Stationarity is dead: whither water management? Science 319(5863): 573–574

Numaguti A, Sugata S, Takahashi M, Nakajima T, Sumi A (1997) Study on the climate system and mass transport by a climate model. In: National Institute for Environmental Studies, Environment Agency of Japan (eds) CGER's supercomputer monograph report, vol 3

Oki T, Agata Y, Kanae S, Saruhashi T, Yang D, Musiake K (2001) Global assessment of current water resources using total runoff integrating pathways. Hydrol Sci J 46:983–996

Oki T, Entekhabi D, Harrold TI (2004) The global water cycle. In: Sparks RSJ, Hawkesworth CJ (eds) The state of the planet: frontiers and challenges in geophysics. American Geophysical Union, Washington, D. C

Oki T, Kanae S (2004) Virtual water trade and world water resources. Water Sci Technol 49 (7):203–209

Oki T, Kanae S (2006) Global hydrological cycles and world water resources. Science 313:1068–1072

Pales JC, Keeling CD (1965) The concentration of atmospheric carbon dioxide in Hawaii. J Geophys Res 70(24):6053–6076

Tyndall J (1861) On the absorption and radiation of heat by gases and vapours, and on the physical connection of radiation, absorption, and conduction. Phil Mag 22(ser. 4):159–194

Utsumi N, Seto S, Kanae S, Maeda E, Oki T (2011) Does higher surface air temperature intensify extreme precipitation? Geophys Res Lett 38:L16708. doi:10.1029/2011GL048426

Vörösmarty CJ, Green P, Salisbury J, Lammers RB (2000) Global water resources: vulnerability from climate change and population growth. Science 289(5477):284–288

Vörösmarty CJ, McIntyre PB, Gessner MO, Dudgeon D, Prusevich A, Green P, Glidden S, Bunn SE, Sullivan CA, Reidy Liermann C, Davies PM (2010) Global threats to human water security and river biodiversity. Nature 467:555–561

Adapting to Climate Variability and Change in India

J. Bird, S. Roy, T. Shah, P. Aggarwal, V. Smakhtin, G. Amarnath, U.A. Amarasinghe, P. Pavelic and P.G. McCornick

Abstract Responding to rainfall variability has always been one of the most critical risks facing farmers. It is also an integral part of the job of water managers, whether it be designing interventions for flood management, improving the reliability of water supply for irrigation or advising on priorities during drought conditions. The conventional tools and approaches employed are no longer sufficient to manage the increasing uncertainty and incidence of extreme climate events, and the consequent effects these have on human vulnerability and food security. To be effective, the technological advances need to be matched with physical, institutional and management innovations that transcend sectors, and place adaptation and responsiveness to variability at the centre of the approach. This chapter examines a number of these challenges and possible solutions at a range of scales, from 'climate-smart villages' to national policy, with a focus on Asia and India, in particular.

1 Introduction

Climate variability already poses a significant threat to humanity, with the poor and most vulnerable at greatest risk. It is expected that such groups will become more exposed to climate shocks as a result of climate change. According to the Fifth Assessment Report of the Intergovernmental Panel on Climate Change (IPCC),

J. Bird (✉) · S. Roy · V. Smakhtin · G. Amarnath · U.A. Amarasinghe · P.G. McCornick
International Water Management Institute (IWMI), Colombo, Sri Lanka
e-mail: j.bird@cgiar.org

T. Shah
International Water Management Institute (IWMI), Anand, India

P. Aggarwal
International Water Management Institute (IWMI), New Delhi, India

P. Pavelic
International Water Management Institute (IWMI), Vientiane, Lao PDR

A.K. Biswas and C. Tortajada (eds.), *Water Security, Climate Change and Sustainable Development*, Water Resources Development and Management, DOI 10.1007/978-981-287-976-9_4

climate change will impact every economic and natural resource sector (agriculture, industry, water, soils, coastal, biodiversity, ecosystems, etc.), and negatively affect people's livelihoods, health and well-being. Rising temperatures and variability in rainfall will continue to result in the increased frequency and intensity of natural disasters, such as floods and droughts. The impact of such variability could undermine the water security of over one billion people, globally, and put the stability of food systems at risk, especially in parts of Asia and sub-Saharan Africa that are vulnerable to hunger and malnutrition (IPCC 2014).

South Asia, India in particular, is at extreme high risk of the possible impacts of climate change. India's economy is closely linked to its natural resource base, with over 700 million people in the country dependent on climate-sensitive livelihood sectors, such as agriculture, water and forestry, which are further threatened by the impact of climate change (GoI 2008). These threats, however, will not be felt uniformly. Availability of renewable freshwater already varies considerably across the country. With growing hydro-climatic variability, much of India is expected to receive a larger volume of its precipitation during extreme rainfall events, such as in western and peninsular India, while other areas will experience fewer very wet days, more dry days and a heightened frequency of extreme weather events.

Increasing climate variability will further complicate farmers' planning and decision making, and the guidance they seek from government agencies within the water and agriculture sectors. As one of the largest economic drivers in the country and by far the largest water user, India's agriculture sector will need to continually adapt to this new reality. Yet, the conventional tools and approaches employed are no longer sufficient to deal with both the significant uncertainties of climate variability and the expectations of stakeholders in an increasingly connected world.

Numerous options are emerging that are promising from an overall water resources perspective, and also provide greater resilience for those exposed to increasing climate-related risks. For example, the potential of integrated management of underground and surface water storage, and using aquifer recharge for multiple beneficial uses, including irrigation, presents innovative adaption options from farm plot to the basin scale. This builds on earlier research conducted by the International Water Management Institute (IWMI) on small and large surface storage (McCartney and Smakhtin 2010), and shallow and Managed Aquifer Recharge (MAR) at the basin and sub-basin scale. Their work demonstrated the importance of examining a range of storage options for both surface water and groundwater that can manage the variability and provide reliable water supplies throughout the year, thereby ensuring water security for communities with investments, and operation and maintenance levels commensurate with their needs and management capacities. The growing use and accessibility of 'big data' (e.g., real-time satellite and weather information) to develop comprehensive flood and drought mitigation plans, and disseminate information rapidly, can enable water resource and disaster managers, communities and farmers to better manage risks related to climate variability, and its impact on agriculture and food security.

Going beyond climate adaptation to mitigation through technical innovations on the ground, for example, using renewable solar energy for pumping water for irrigation, potentially offers the opportunity to reduce India's carbon emissions by around 8 % per year (Nelson et al. 2009). Also, if appropriate incentive structures are in place, there is potential to replace the present over-abstraction of groundwater and reduce the demand for electricity. This and other innovative incentives, technologies and approaches are being piloted in a network of 'climate-smart villages' (CSVs) in several states in India and elsewhere. Farmers adopt a portfolio of climate-smart interventions that include a range of practices and technologies suited to the local agricultural systems and needs.

This chapter presents research conducted by IWMI and its partners under the CGIAR Research Programs on Water, Land and Ecosystems (WLE), and Climate Change, Agriculture and Food Security (CCAFS), seeking innovative, viable approaches for building India's climate resilience in the agriculture sector. It examines a number of challenges posed by increasing climate variability and presents possible solutions at a range of scales, from CSVs to policy interventions at basin and national levels. This chapter builds on a presentation made by Jeremy Bird, Director General, IWMI, during the Seminar on Water Security, Climate Change and Sustainable Development at the recently concluded "Vibrant Gujarat Summit 2015" held in Ahmedabad, India, on January 11–13, 2015.

2 Aquifer Storage and Recharge from Farm to Basin Scale—An Adaptation Response to Heightened Hydro-Climatic Variability

With growing hydro-climatic variability, India's *Rabi* and summer crops will experience higher evapotranspiration (ET) demand (Shah 2009a). According to IPCC (2001), peninsular India is expected to receive 5–10 % higher precipitation in absolute terms, although increasingly during 'very wet days'. Increased frequency of extremely wet rainy seasons (Gosain and Rao 2007) is also likely to mean increased runoff. According to Milly et al. (2008), compared to the period 1900–1970, most of India is likely to experience a 5–20 % increase in annual runoff during the period 2041–2060. As a result, surface storages will receive larger runoff but, thanks to higher temperatures, will also suffer from higher evaporation losses. Higher ET demand of crops will also mean an increase in seasonal irrigation requirement and a reduced command area (Shah 2009a). Therefore, in the absence of other efficiencies, irrigation from surface storage will suffer some loss of effectiveness because irrigating the same command area would need larger dam storage to provide the same level of water security. Similarly, climate change will have some adverse effects on groundwater storage. Natural recharge to aquifers may decline due to the shorter but heavier runoff events. Despite these factors, if managed appropriately, aquifer storage and groundwater irrigation could provide an opportunity to become more significant as a climate adaptation mechanism to stabilize Indian agriculture, and support water security across sectors.

Aquifers respond to droughts and climate fluctuations slower than surface storage. As a result, aquifers act more as a resilient buffer during dry spells. This is one reason why India has experienced an explosive growth in groundwater demand during recent decades, and also why this demand will expand further in the wake of climate change. For millennia, groundwater wells have been the principal weapon Indian farmers have used to cope with droughts (Shah 2009b). Indeed, the digging of wells has tended to peak during years of droughts. This trend continues today and will likely increase with heightened hydro-climatic variability. Ironically, the over-abstraction of groundwater due to perverse energy pricing has significantly stressed and reduced the resilience of key groundwater resources, thereby decreasing its effectiveness in adapting agriculture and other sectors to the vagaries of climate change. Therefore, a change of focus is needed to shift irrigation investments towards a more sustainable pathway, away from surface storages and excessive use of groundwater to systematically managing groundwater abstractions and replenishment through MAR. This approach is especially pertinent to India's west-to-south corridor where years of groundwater over-exploitation have left aquifer systems depleted (Fig. 1).

Historically, India has emphasized surface storage and gravity-flow irrigation to ensure food security, which was followed by widespread development of groundwater over the past three decades. However, climate change raises new questions about the reliability of storage systems and the need for a more integrated strategy. Table 1 compares four storage alternatives through the lens of eight criteria using a ten-point scale that assigns a score of up to five points for positive attributes (benefits) and up to minus five points for the negatives (costs, disbenefits). The four storage alternatives compared are as follows:

- **Small surface storages** are advocated by environmental and civil society groups because they are close to the point of use, have short canals and can be effectively managed at the community level. India's age-old traditional water harvesting structures, such as tanks in South and eastern India, *Ahar-Pyne* systems of southern Bihar, homestead ponds of West Bengal and North Bihar, and *johads* of Rajasthan, represent this class (Choppra 2005).
- **Large surface storages** are generally constructed and operated for multiple uses by government agencies at topographically and hydrologically suitable locations. Water from these storage systems is transported through an extensive network of surface canals. These systems generally have an agency to oversee the operations, and are socially and institutionally complicated to manage effectively. However, such systems represent a very important asset in relation to water and food security.
- **Shallow aquifer storage** has supported the groundwater boom that India has experienced through the investments of millions of small farmers, although without any demand-side management or a systematic strategy for enhancing aquifer recharge, which has raised sustainability concerns.
- **Managed Aquifer Recharge** (MAR) represents an option that is not practiced widely in India at present. However, with a major shift in thinking, and in combination with proactive demand management, it could sustain groundwater resources even to meet increasing demands.

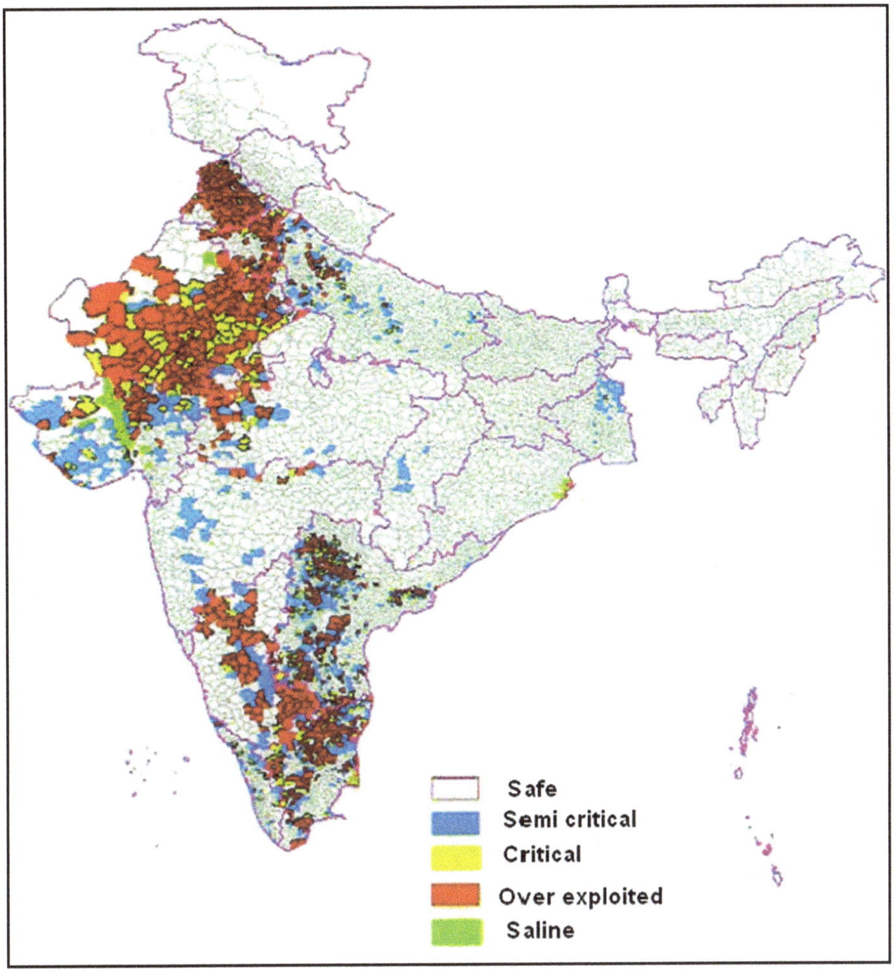

Fig. 1 Groundwater-stressed areas of India. *Source* Shah (2009a)

What is important in the scoring assigned to each storage option is not the absolute numbers, but the relative comparison. Rows four and eight (Table 1) indicate costs or disadvantages of different storage options, whereas other rows suggest benefits on different scales for other criteria. Of the costs and benefits, some options, such as the level of water control, are private in nature and drive the choices of individual farmers. Others are 'public' (or social) in nature; for instance, the carbon footprint of alternative storage systems may not directly influence the individual decisions of farmers, but has to be factored into national planning considerations.

Since the 1970s, the advantages of groundwater irrigation in terms of location, ease of access and flexibility of control (reflected in Table 1, rows one, two and three) have driven India's groundwater boom. The resilience of groundwater against

Table 1 Climate change and water storage alternatives

	Measurable criteria	Small surface storage	Large surface storage	Shallow aquifer storage	Managed aquifer recharge
1	Water where needed	+++	++	++++	+++++
2	Water when needed	+	++	++++	+++++
3	Level of water control	+	++	++++	+++++
4	Non-beneficial losses, e.g., evaporation	–	– –	–	–
5	Protection against a single annual drought	+	++	+++	+++++
6	Protection against successive droughts	–	+	++	++++
7	Ease of recovery during monsoon	+++++	++++	++	+++
8	Carbon footprint	–	– –	– – – – –	– – –

Source Adapted from Shah (2009a)

dry spells and droughts (Table 1, rows five and six) is also important. Surface storage has, in contrast, fared poorly on these counts. These benefits will become more valued as climate change heightens the hydrological variability. From an overall water resources viewpoint, aquifer storage also has the advantage of minimum non-beneficial evaporation (Table 1, row four). For a mostly semi-arid country, where surface reservoirs can lose 3 m or more of their storage every year to surface evaporation, this is a significant gain. The major socioeconomic disadvantages of a heavy dependence on groundwater are as follows:

a. Aquifers are slow to recharge, and the hard-rock aquifers that underlie 65 % of India have limited storage.
b. While gravity-flow irrigation from canals needs little or no energy, groundwater irrigation is energy-intensive.
c. Since the bulk of groundwater pumping uses diesel fuel or electricity generated from coal, India's transition from surface gravity irrigation systems to pumped irrigation has created a massive carbon footprint. These electric and diesel pumps represent a total installed capacity of 130 gigawatts (GW). India's 11 million electric tube wells use an estimated 117 billion units of electricity every year and emit 110 million metric tonnes (MMt) of carbon dioxide (CO_2) into the atmosphere. Also, nine million diesel pumps use an estimated 7 billion litres of diesel for irrigation every year and emit an additional 20 MMt of CO_2. Together, this amounts to nearly 6.6 % of India's total annual CO_2 emissions of 1970 MMt. However, Nelson et al. (2009: Table 11) estimated the CO_2 equivalent of greenhouse gas (GHG) emissions from India's groundwater economy to be much higher at 186.4 MMt, which is over 8 % of India's total annual CO_2 emissions.

These downsides could be prevented by an effective MAR programme that replenishes the depleted aquifer systems of western and southern India on an annual basis, mitigates flood extremes (see Sect. 4.1) and utilizes solar power as the energy source (see Sect. 3).

2.1 Managed Aquifer Recharge: Conjunctive Use of Surface Water and Groundwater for Basin-Wide Climate Change Adaptation

According to EM-DAT (2014), about three billion people, globally, are affected by floods and droughts. Between 1980 and 2008, about 25,000 deaths, on average, every year were associated with floods and droughts worldwide, with economic losses of USD 165 billion. Flooding creates enormous suffering, displacement and financial losses to the poorest and most vulnerable communities in states of India, such as Gujarat, as well as Punjab, Haryana, Uttar Pradesh, Bihar, Odisha and others. The seasonal contrasts in water availability are extreme. Based on current trends, climate change will likely lead to even more reliance on groundwater. So, if floodwaters could be used more extensively to recharge aquifers, both floods and droughts could be addressed simultaneously.

The practice of harvesting surface water and storing it underground in aquifers through MAR has been around on the Indian subcontinent in traditional forms for millennia and in more modern forms since around the 1960s. MAR seeks to utilize runoff water more effectively through interventions that facilitate groundwater storage, such as infiltration wells, percolation tanks, subsurface barriers, spreading ponds, check dams, and recharge pits and shafts. MAR effectively offers a means of augmenting groundwater supplies to secure and enhance water supplies for drinking purposes and irrigation; improve groundwater quality or prevent saltwater intrusion; reduce evaporation of stored surface water; conservation or disposal of floodwaters; storage of water to reduce pumping and piping costs; temporary regulation of groundwater abstraction; and maintain environmental flows and groundwater-dependent ecosystems. Thus, the associated socioeconomic benefits of MAR include increasing groundwater availability to help farmers cope with the rainfall variability exacerbated by climate change; and mitigation of flood impacts. Other, newer modalities of MAR, including Underground Taming of Floods for Irrigation (UTFI) (Pavelic et al. 2012) and subsurface storage (SSS) (Amarasinghe et al. 2015; Muthuwatta et al. in prep.), address both seasonal floods and droughts conjunctively, as discussed later.

In India, available water resources have been heavily exploited due to growing urbanization, and the demand for increased agricultural and industrial production. The country possesses 18 % of the world's population, i.e., 1.25 billion people, but accounts for only 4 % of the world's water resources. With the population increasing rapidly, water scarcity issues have increased multifold and per capita water availability has been falling. In urban areas, the amount of rainwater infiltrating into the ground has decreased due to the construction of buildings, roads and other hard surfaces. Frequent failure of monsoons and limited surface water resources have led to the increased dependence on groundwater. Large sections of the rural population of India that depend directly on groundwater for drinking purposes (more than 85 %) and irrigation (more than 60 %) (World Bank 2010) are at risk due to depletion of the resource and water quality degradation.

As of 2011, one-third of the administrative blocks of groundwater resources in the country were being utilized at levels that are not considered 'safe'. MAR has been looked upon as a solution in the hard-rock areas of western and southern India, where there is a declining trend in groundwater level in over-exploited aquifers. MAR is sometimes seen as a panacea for groundwater over-exploitation and more appealing to implement than demand management measures, which requires constraining the rates of abstraction, and involves related political and institutional challenges. Whilst MAR is practiced globally to varying extents, India is an international leader with 4 km^3 of water stored underground, annually. The Central Ground Water Board (CGWB) of India, through its current Master Plan, aims to recharge much of the 85 km^3 of surplus runoff per year with the construction of 11 million structures at an estimated capital cost of around USD 12,000 million (CGWB 2013).

MAR interventions are widely adopted and supported by local communities, and state and central governments. Large regional programs are being implemented across the country under various national government flagship programs, such as the Integrated Watershed Management Program (IWMP), the Mahatma Gandhi National Rural Employment Guarantee Act (MGNREGA), Bharat Nirman and others. Gujarat has invested heavily in MAR over the past two decades, and is acclaimed as a success story by decision makers and perhaps a representative model for the rest of the country (Shah 2009b).

While the literature on MAR in India is vast, as indicated by review papers such as those by Gale et al. (2002) and Sakthivadivel (2007), the science to support the high expectations of MAR and justify the financial investments being made in it is more limited. The available scientific evidence is mixed in terms of the impacts of MAR. Prathapar et al. (2012) argued that there is overwhelming evidence that most MAR structures are poorly designed and located, and therefore provide commensurately poor investment of valuable human and financial resources. Implementation of many MAR projects rely on very little scientific and technical input, and there is a lack of capacity in many government and nongovernmental organizations (NGOs) to implement such projects effectively (Reddy and Syme 2015). Kumar et al. (2008) also took a pessimistic view of MAR after considering long-term institutional support, maintenance issues, externalities, water pollution, cost-benefit ratios and downstream trade-offs. All this points to the need for a more rigorous monitoring and evaluation programme feeding back into project design.

MAR is an appealing and necessary response to the escalating problem of groundwater over-exploitation in India. The stakes are high, for example, with the economic value of groundwater irrigation estimated to be about USD 8 billion per year (Shah 2008). Ensuring that resources spent on MAR achieve positive net benefits for agricultural production and minimize the negative externalities is a key challenge.

This can only be achieved, if appropriate attention is given to the institutional strengthening of implementing agencies, scientific planning for siting and design, and ongoing resource allocations which are dedicated to addressing operational issues, such as de-silting and maintenance of structures, over the long term.

In exploring the concept of MAR on a landscape or basin scale, IWMI has been considering two distinct but interconnected approaches which have been referred to above: (i) UTFI, and (ii) creating more SSS prior to the monsoon.

1. UTFI involves recharging depleted aquifers within the upper reaches of basins by capturing and, where necessary, diverting excess wet-season flows. This alleviates downstream flooding, and provides a resilient groundwater supply for dry-season irrigation and for use in other sectors (Pavelic et al. 2012). This process requires the participation and collective action of local institutions (farmers, local government) in the upstream areas, with suitable incentive mechanisms linked to institutions in more populated areas downstream. While there are similarities between UTFI and existing general watershed development and aquifer recharge programs in India, existing programs have historically targeted the most water-scarce, more arid settings at the micro- or meso-scales, with limited attention being given to more humid regions where extreme seasonal excesses and scarcity are still problematic.

   Conditions required for successful implementation of UTFI have been analyzed and appear to be favourable in significant parts of the upper, middle and lower reaches of the Ganges River Basin (Fig. 2). Planning is underway to implement UTFI at the pilot scale, which aims to demonstrate the benefits, costs and trade-offs of the approach, and provide the evidence base and strategies needed to secure policy change and investment at scale.
2. SSS is an approach that creates additional storage capacity underground by increasing the volume of water pumped out prior to the onset of the flood season. It has the potential to provide multiple ecosystem services and benefits beyond irrigation. SSS buffers water variability by providing the supplies needed to sustainably intensify and improve productivity in agriculture, and also for the domestic and industrial sectors. It has the potential to increase the irrigated area by 40 % in the Ganges River Basin, and meet the domestic and industrial water demand of large cities. SSS also avoids numerous social and environmental costs associated with surface storage development, such as the rapid reduction of surface reservoir storage due to the high sediment loads emanating from the young Himalayan Mountains during flood events. Regulation of flow through SSS can also help alleviate the social impacts of floods and droughts, especially on women and children who are often the hardest hit by extreme climatic events.

   The present surface storage capacity within the Ganges River Basin is only 9 % of the total annual runoff of 525 billion cubic metres (Bm^3) (equivalent to 88 m^3/person), which is very low compared to countries that have made major investments in infrastructure, such as 5,961 m^3/person in the United States of America and 2,486 m^3/person in China. A range of reasons explain this difference, including topography, financing, geopolitics and transboundary

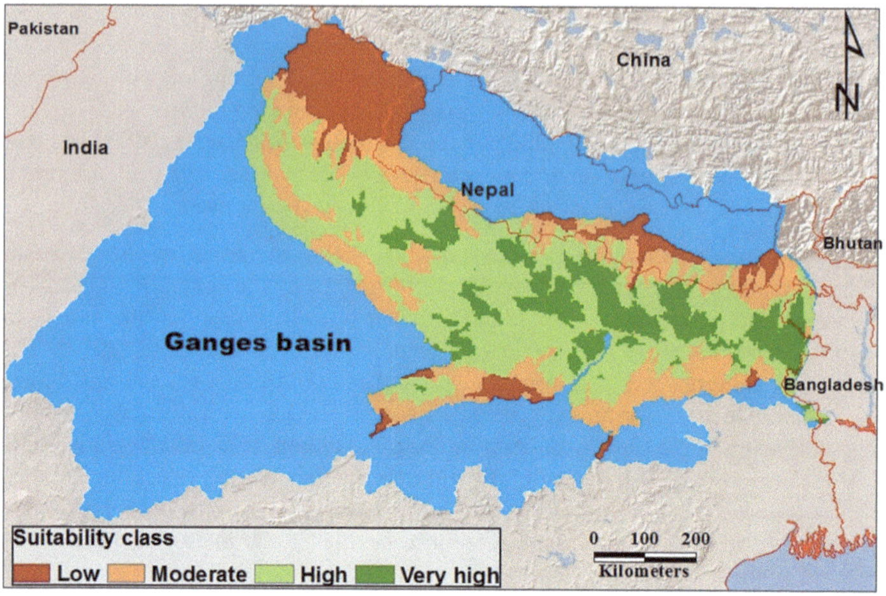

Fig. 2 Preliminary assessment of the suitability of the interior part of the Ganges Basin to UTFI based on 30+ biophysical and economic indicators. *Source* Karthikeyan and Pavelic (in prep.)

relationships, which have constrained surface storage development in the densely populated and poor regions of the Ganges River Basin.

Looking underground and creating more SSS is a potential solution to the water storage dilemma. The approach entails pumping additional volumes of groundwater out of the aquifers before the onset of the monsoon, and using this water for additional irrigation during the dry season or securing it for the domestic and industrial sectors during the period when water is normally critically scarce. This is termed ‘preparatory pumping’. The important part of this approach is creating additional, or meeting unsatisfied existing, water demand, which is applicable in both arid and humid areas of the Ganges, and potentially in other river basins. Subsequent recharge through monsoon rainfall and runoff replenishes the aquifers, thus ensuring sustainability of the enhanced benefits. This concept was first proposed as the “Ganges Water Machine,” as a grand challenge to capture a larger portion of the monsoon rains for use in the dry season (Revelle and Lakshminarayana 1975). Although some local examples of SSS are in place, it has not been implemented yet at the scale required because of the enormity of the challenge (http://wle.cgiar.org/blogs/2013/05/21/the-ganges-water-machine-a-sexy-title-or-the-concept-of-global-relevance/).

At present, river flow in the Ganges River Basin is substantially higher than consumptive water use (CWU) during the monsoon season (June to October) and lower in the non-monsoon season (Fig. 3). The demand for water, especially for irrigation, is increasing rapidly with the rising population and growing

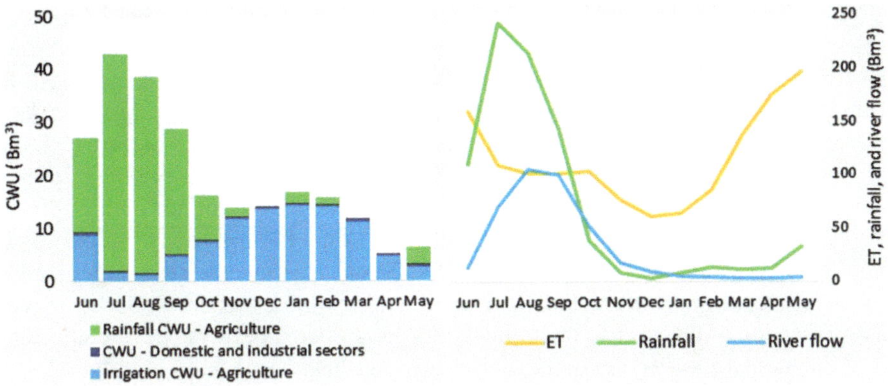

Fig. 3 Monthly evapotranspiration (ET), rainfall and river flow in the Ganges Basin. *Source* CWU—authors' estimates; Rainfall—Indian Meteorological Organization. The river flows are based on the daily observed data at the Harding Bridge gauging station in Bangladesh just south of the Indian border

economies. Yet, without adequate storage, any further increase in CWU, particularly during the *Rabi* (November to March) and hot weather (March to May) seasons, could severely stress river flows and impact riparian ecosystems.

There is potential for SSS in many other large river basins in India. Basins such as Narmada, Mahanadi, Godavari and Brahmaputra, with a combined population of 690 million, all have large volumes of surface runoff, with more than 80 % of the 1280 Bm^3 of runoff in the basins generated during the monsoon season. Water stress is a recurrent phenomenon at other times. As with the Ganges, only a small fraction of the runoff can be potentially utilizable with surface storage. Assessments of current water use, future water demand, surface water and groundwater hydrology, environmental flows, socioeconomic conditions and institutional frameworks would be required in order to put SSS into practice in these basins.

3 Solar Irrigation Pumps—A Solution for Climate Change Adaptation and Mitigation

By providing on-demand irrigation, groundwater use in agriculture has buffered some of the downside impacts of increasing hydro-climatic variability, and provided the water security necessary for investing in high-value crops. However, India's groundwater socio-ecosystem is facing several critical problems. The most important of these issues, as discussed above, is the persistent depletion of alluvial as well as hard-rock aquifers throughout the western Indian corridor from Punjab to Tamil Nadu. The MAR approaches described above can be even more effective if the perverse incentives arising from farm power subsidies, which have distorted farmers' demand for groundwater, are eliminated.

While groundwater levels continue to drop, it is India's electricity industry which suffers the most as a result of perverse farm power subsidies. In theory, state governments are obligated to reimburse the power distribution companies. In reality, however, no state government fully compensates its power companies. As a result, power utilities in India incur annual losses in the order of nearly USD 10 billion/year[1] (Shah et al. 2014). It is not surprising, therefore, that power companies treat farmers as their 'loss leaders', offering them a poor quality power supply mostly during the night when other demand for electricity is low. This situation has led to various outcomes in different regions. In eastern India, for instance, poor farmers are unable to benefit from abundant groundwater resources due to the lack of electricity. In contrast, in western parts, perverse power subsidies have threatened the sustainability of groundwater-based irrigated agriculture.

The arrival of solar energy promises to resolve these problems. However, the present policy environment in India, which aims to achieve its solar energy target primarily by offering generation-linked incentives to large, megawatt-scale concentrated solar or solar thermal plants, needs to be reconsidered. Some have been arguing in favour of distributed generation through rooftop photovoltaic (PV) plants. Yet, others are arguing for micro-grids to light up remote rural communities.

As of now, the promotion of solar pumps in India is a small part of the more ambitious Jawaharlal Nehru National Solar Mission (JNNSM), which is mandated to raise solar power generation in the country from approximately 3 GW, at present, to 100 GW by 2022. Each state has its own solar pump promotion programme with a capital cost subsidy of 70–95 %. State governments are offering solar pumps to redeem pending applications from farmers for grid power connections on tube wells. This means that, in the western and southern states of India, solar pumps that offer free, un-interrupted daytime energy for pumping groundwater are in addition to the grid-connected pumps that are subject to perverse electricity subsidies. Continuing this solar pump promotion policy will only further exacerbate India's groundwater depletion problem. Moreover, as solar panels become more affordable, the number of solar pumps in use will multiply even without government subsidies, and this will pose a serious threat to the sustainability of the country's groundwater resources.

To transform this perverse outcome into a virtuous one, IWMI researchers have argued that India should incentivize farmers to 'grow' Solar Power as a Remunerative Crop (SPaRC). In particular, this would include the creation and operation of a solar pump promotion fund that will:

a. provide a subsidized solar pump to tube well owners in lieu of surrendering their grid power connection;
b. ensure net metering of such solar pumps and connect them with the grid;
c. offer all the solar pump irrigators a guaranteed buyback of surplus energy at an attractive price comparable to what is offered by megawatt-scale power plants, so as to provide an incentive for more efficient water management;

---

[1]USD 1 = INR 64 (as at July 2015).

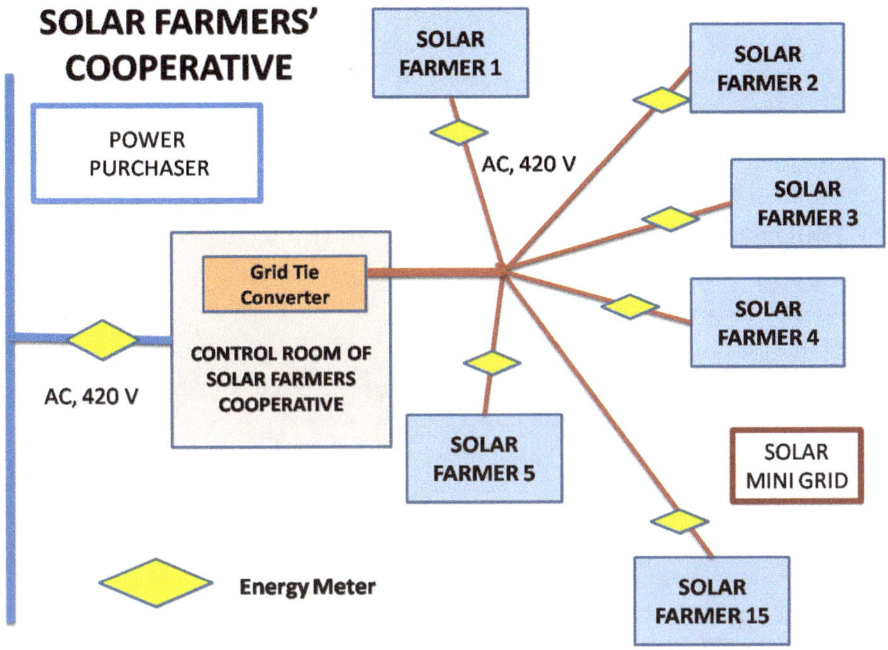

Fig. 4 Redesigning the solar mission as a 'cash crop' opportunity

d. provide power distribution companies with an incentive for every unit of solar power they purchase from solar pump irrigators;
e. provide farmers with the freedom to expand their solar power generation capacity at their own cost; and
f. organize farmers using solar pumps into a village solar pump irrigators' cooperative enterprise (SPICE) (Fig. 4), which would pull and evacuate all surplus solar power of its members at a single point.

Such a solar pump promotion policy could resolve several problems at the same time. Solar power generation can provide farmers with a lucrative income stream. One hectare of grain production yields a net annual income of approximately USD 800–1,000. One hectare devoted to fruit and vegetable cultivation can yield a net income of USD 2,000–3,200. However, a hectare under solar panels could generate energy with a market value of USD 50,000–120,000 per year. While it is true that most farmers would be unable to mobilize the capital needed for a solar pump of this size, even a 10-kilowatt peak (KWp) solar plant taking up 1 % of a hectare would still yield the equivalent income of a full hectare of crops.

Solar pumps could also be a powerful climate change mitigation strategy. Solarizing India's 20 million irrigation wells could reduce the country's carbon emissions by 8 % per year. Similarly, solarizing the groundwater economy could not only liberate eastern India's agriculture sector from persistent electricity shortages, but it can also free western India's power industry from the deadweight of perverse farm subsidies. Above all, a solar pump economy can replace the present perverse incentive to waste power and water by a virtuous incentive to conserve both. By offering farmers guaranteed buyback of surplus solar power at an attractive price, the strategy that IWMI proposes can raise the opportunity cost of using solar energy for pumping groundwater.

4 Understanding and Quantification of the Risks of Floods and Droughts: The Role of 'Big Data' and New Data Acquisition Technologies

Now, more than ever, researchers have an unprecedented wealth of data at their fingertips from continuously orbiting satellites, weather monitoring instruments, ecological observatories, model simulations and forecasts (Sellars et al. 2013). With just a fast internet connection, scientists and engineers can access atmospheric and oceanic gridded data and time series observations from around the world, minute-by-minute conditions of the near-Earth space environment, and other data streams that provide information on events across local, regional and global scales. These datasets have become essential for monitoring and early warning of floods and droughts, and understanding the societal impacts.

The role of the 'big' Earth data initiative will have wider applications, such as climate change and risk modelling, weather insurance, managing aquifer recharge and revealing critical insights for strengthening resilience, including in the agriculture sector—something that was never possible before. These data allow water resource and disaster managers in South Asia to better manage risks related to climate variability, and its impact on agriculture and food security, as depicted in Fig. 5.

4.1 Systems to Monitor and Manage Floods and Droughts

At a subcontinental scale, IWMI has developed flood and drought monitoring (Figs. 6 and 7) approaches and tools that allow spatial (500 m) and temporal (8-day interval) mapping of floods and droughts. This is done using the National Aeronautics and Space Administration's (NASA's) Earth Observing (EO) System Data, namely the Surface Reflectance and Land Surface Temperature to calculate land and water indices, including the Normalized Difference Vegetation Index (NDVI), Land Surface Water Index (LSWI) and the Enhanced Vegetation Index (EVI), to identify the extent of flooding and droughts for several countries in Asia and Africa (Amarnath et al. 2012; Amarnath 2014). The products can also be used proactively to study flood vulnerability, areas prone to waterlogging and the impact on critical agricultural production zones. It provides a planning and response tool to help reduce vulnerability, increase people's resilience to climatic shocks and stresses, and manage floods and droughts more holistically from local to basin and regional scales.

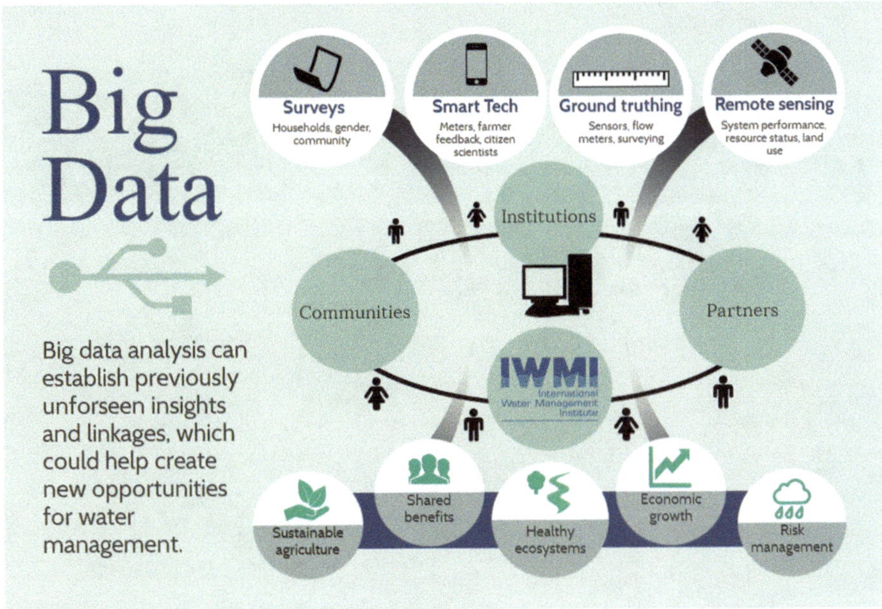

Fig. 5 Big data helps improve global water, food security and climate risk management. *Source* IWMI

The application of drought monitoring from Earth observation data can address drought risk evaluation, characterization of drought episodes and development of drought risk indicators, and selection and prioritization of measures to alleviate the effects and assess the impacts of droughts to better inform the decision-making process. The timely and accurate monitoring of drought extent could be important for agricultural drought risk management, water management and irrigation scheduling for crop production. For example, semi-arid regions in India have been historically impacted by water scarcity and droughts with consequent socioeconomic and environmental impacts. IWMI has started a drought monitoring programme for South Asia, and the products developed will support agricultural drought risk management, and provide timely information to water resource authorities and disaster agencies.

5 Climate-Smart Villages: Maximizing the Benefits of Water-Smart Interventions

While water-based strategies, as outlined above, are crucial in managing climatic risks, it is necessary that other critical factors are also considered for enhancing crop production. These factors include technological, institutional and policy options, such as changes in agronomic practices, weather-based index insurance, weather-based

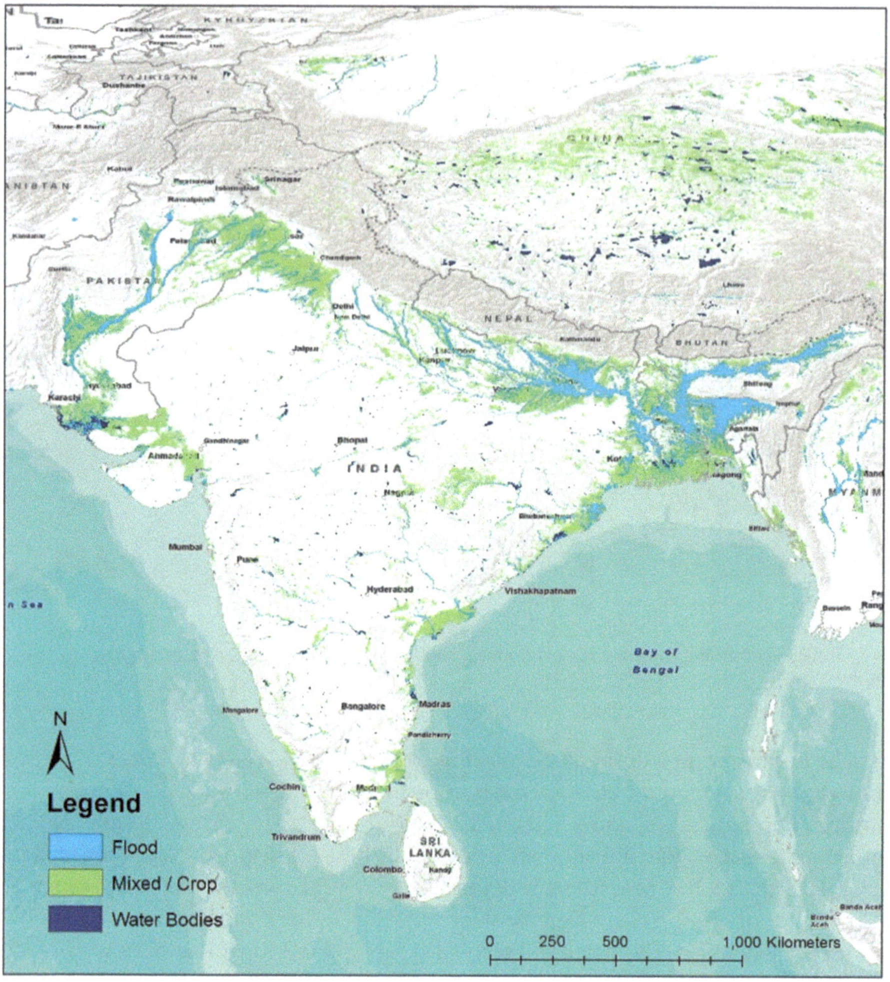

Fig. 6 Maximum flood inundation extent derived using Moderate-Resolution Imaging Spectroradiometer (MODIS) time-series images from 2000 to 2012 covering South Asia. *Source* Amarnath et al. (2012)

agro-advisories and stress-tolerant seed varieties, and empowerment of self-help groups or water policy changes. These practices and technologies, individually, have proved to be useful in increasing production and incomes, and lowering GHG emissions. Each of them, however, comes with varying costs, benefits and economic impacts. Their implementation requires appropriate investment decisions in both on-farm capital and for wider agricultural outreach programs. In all these cases, especially water and energy, interventions can be at the individual farmer level (e.g., on-farm water management, biogas plants and solar pumps), village community scale (e.g., rainwater harvesting, solar irrigation micro-grids, laser levelling), or

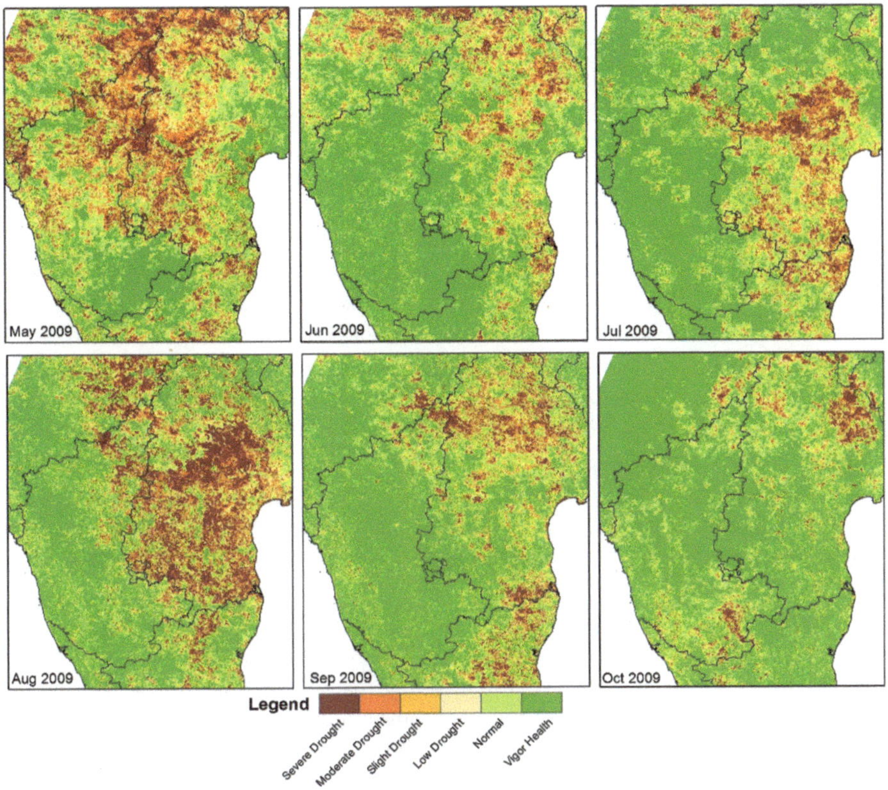

Fig. 7 Seasonal variation of drought severity captured using MODIS time-series images for the drought event in 2009 covering South India. *Source* IWMI

agroecological or sub-national level (e.g., food and seed banks, aquifer recharge, solar pumps).

Climate-smart agriculture (CSA) is an approach which strives to sustainably increase productivity and incomes, build resilience and adaptive capacity to climate change and, where possible, reduce/remove GHG emissions from agriculture (FAO 2010). Therefore, this approach is increasingly gaining global attention. It works to enhance achievement of food security and development goals at different spatial and temporal scales—from farm to landscape, local to global, and over short and long time horizons. With its explicit emphasis on the management of climatic risks that are increasing in frequency and intensity, information flows and local institutions support adaptive capacity. The CSA approach is highly complementary with the 'sustainable intensification' approach (Campbell et al. 2014). CSA differs from 'business-as-usual' approaches by emphasizing the capacity to implement context-specific solutions, building evidence for scaling-out, and increasing local institutional effectiveness supported by innovative policy and financing actions (Lipper et al. 2014).

There are many projects that are testing or promoting CSA, but few have shown widespread uptake of the approach (FAO 2015; Neate 2013). We illustrate below some specific interventions in India that have reached a particular scale and promote some aspects of CSA.

Crop insurance: Risk transfer approaches, e.g., crop yield insurance and weather derivatives, are increasingly being used as viable solutions for managing risks associated with climatic variability. Today, almost 30 million farmers in India are covered by some kind of crop insurance. This includes almost 13 million farmers covered by weather-based crop insurance schemes. In principle, such subsidized schemes are supposed to protect farmers from climatic extremes, and also encourage them to practice high-yielding agriculture. Despite this, there is large-scale dissatisfaction among farmers with insurance products, because of residual, spatial and temporal basis risk, delays in the settlement of claims and limited infrastructure to measure reference weather. Insurance schemes are now being evolved to overcome these limitations by increasing the density of weather stations, developing improved agroecological region-specific weather triggers, bundling insurance with the rest of the risk management portfolio, such as water management and the use of drought-tolerant seeds, and the use of remote sensing for village-level yield mapping.

Weather-linked agro-advisories: Farmers and other stakeholders can considerably increase the efficiency of agriculture, if actionable knowledge based on seasonal and short-term weather forecasts is made available to them in time. Increasing availability of information and communications technology (ICT) tools and mobile phones has made this possible all over the world. Currently, there are several such initiatives to make this information available to farmers, which are at different stages of development in many countries. The Indian Meteorological Department has been providing such information at district and state level to farmers, industry and the government for some time. District-specific, 5-day weather forecasts, in quantitative terms, are disseminated to more than five million farmers today through partnerships with ICT companies. The information disseminated includes rainfall, cloud, maximum/minimum temperature, wind speed/direction and relative humidity, and also forewarning of hazardous weather events likely to cause stress on standing crops and suggestions to protect these crops from such events. Value-added information such as market prices and health is also provided. There is a need for an independent critical evaluation of such services to understand their limitations and to also identify opportunities for their expansion.

Climate-smart villages: In order to build evidence on the benefits of water and energy management together with other CSA interventions, CCAFS and IWMI are working with national programs and partnering with rural communities to develop CSVs. Built on a community approach, researchers, local partners, farmer groups and policymakers collaborate to select the most appropriate technological and institutional practices that can increase productivity, enhance adaptive practices and lower GHG emissions. These include practices and technologies that are 'smart' in weather, water, carbon, energy, nutrient and knowledge management (Aggarwal et al. 2013). Interventions are specific to each village, but the concept can be applied in any

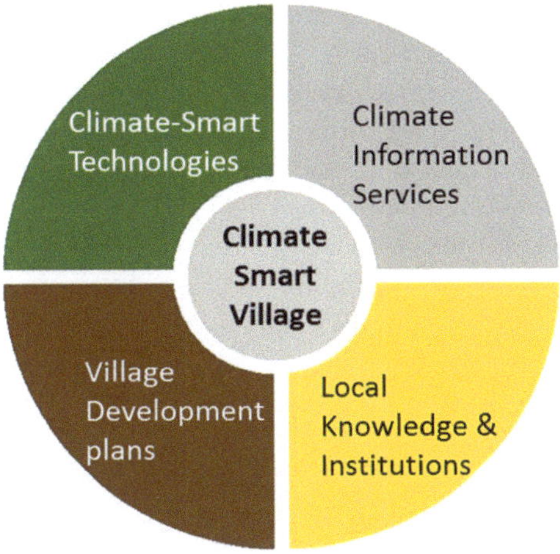

Fig. 8 Climate-smart village model

region under the right circumstances. The key focus of CSVs is to enhance climate literacy among farmers and local stakeholders, and develop climate-resilient agricultural systems by linking to existing village development schemes and investments. Climate information services are an important part of the project (Fig. 8).

In India, CSVs are being piloted in 75 villages in Haryana, Punjab and Bihar. Recent studies indicate that crop yield and net return gradually increased after the second year when conservation agriculture alone was practiced in the rice-wheat systems in Bihar (Jat et al. 2014). Initial modelling results also indicate that there is scope for a significant increase in production and farmers' income, climate resilience, adaptation to longer-term climate change and, to some extent, mitigation of GHG emissions over the long-term when efficient water management practices were combined with good seed varieties, nutrient management and crop insurance.

6 Conclusion

Increasing climate variability due to climate change adds additional levels of uncertainty to decisions, including those faced by farmers, for the pre-season and project planning advice provided by government agencies, and for the longer-term design and management of water resources infrastructure. As variability increases, it also challenges conventional approaches to water management and requires us to promote alternative ideas, for example, on the role of physical interventions to improve the productive use of the water that is available, to make better use of new technologies to plan water management regimes or to intervene in the management of the resource itself, as in the case of groundwater recharge to provide buffer storage. Inevitably, it is the poorest elements of communities that are the most vulnerable to climate variability and least able to finance alternative coping strategies. Hence, climate adaptation is needed at various scales that focus on multiple targets ranging from direct assistance at community level, in support of the livelihoods of vulnerable communities at risk from floods or droughts, to the protection of economic assets that contribute to a country's development growth and consequent capacity to reduce poverty.

In many areas, groundwater management can be a part of the solution as shown through a series of promising examples in this chapter, including storing excess wet-season flows underground (UTFI), and recharging aquifers to provide a storage 'buffer' (SSS) and water for irrigation in the dry seasons and during periods of drought. Groundwater use has been central to the recent expansion in irrigation across South Asia, including India, and also fundamental for urban and industrial development. In some regions, such as the eastern Gangetic Plains, there is scope for even further development. The Groundwater Recharge Master Plan (GRMP) proposed by CGWB reflects belated recognition of the growing criticality of MAR for Indian agriculture. Where surplus runoff is available aplenty, GRMP aims to raise post-monsoon groundwater levels to 3 m below ground level through annual MAR of 36.4 km^3, by constructing, mostly in eastern India, some four million spreading-type recharge structures at a cost of over USD 4 billion. While it is a step in the right direction, GRMP needs to be redesigned as a response to heightened hydro-climatic variability with greater emphasis on arid alluvial aquifers of northwestern India, and besalt and crystalline aquifers of peninsular India; the latter occupies 65 % of India's landmass, have suffered extensive groundwater depletion and are most at risk from climate change. However, in others, such as western and peninsular India, sustainable withdrawal limits have already been exceeded, and measures to improve the productivity of water (that has been espoused for so long) are now becoming essential. Plans such as GRMP need to be reimagined as a participatory programme eliciting the active engagement of farmers and their cooperatives, NGOs and a myriad of other stakeholders. Providing the right incentive frameworks for this to happen is key as has been shown in Gujarat.

Technological advances in solar irrigation pumps are already leading to a major expansion of low-cost, more environmentally friendly tube wells that will increasingly displace pumps driven by diesel or electricity generated from fossil fuels. However, this climate mitigation solution comes with a major risk, if not adequately packaged. The very low cost of solar energy further threatens the sustainability of groundwater levels, and this is a fast-emerging 'nexus' issue at the interface of water, food, energy and the environment. Work being undertaken by IWMI shows that, with the right incentive frameworks to buyback the excess electricity from these pumps, either individually or through a 'solar cooperative', a series of 'wins' can be achieved, including expansion of the irrigated area, decrease in the demand for energy from fossil fuel sources with an associated reduction in carbon emissions, greater efficiency in water management and potentially recovering water levels. Producing a solar crop with benefits to both the farmer and the national economy, together with the conventional agricultural crops, offers a novel way to address this nexus.

More sophisticated and accessible remote sensing data provides new opportunities for planning, management and emergency response in the case of extreme climate events. Adaptation of near real-time drought monitoring technology to the local context is demonstrating a more effective basis for local decisions on crop selection and water management. High resolution analysis of historic flood events offers the opportunity to better assess future risks and target response mechanisms, while rapid analysis of ongoing extreme flow events helps direct emergency relief efforts.

At community level, increasing emphasis is being placed on the awareness of climate resilience through the CSV initiative, in which a range of institutional and technological approaches are focused on both adaptation and mitigation. The emphasis here is to add to, and link with, existing development initiatives whether they be at the farm level (e.g., solar irrigation, on-farm water management, rainwater harvesting) or at a higher agroecological level (e.g., aquifer recharge, seed banks for climate-resilient varieties). A recurring requirement of many of these initiatives is the need for cross-sectoral cooperation in programme development and implementation, something that has often been elusive in previous rural development programs and at higher levels of government policy making. The potential to identify mutually attractive solutions to address the climate challenge may provide the necessary enabling environment for such integrated solutions, for example, the proposed incentive-based approach to influence groundwater management through the 'solar cash crop' concept.

References

Aggarwal P, Zougmoré R, Kinyangi J (2013) Climate-smart villages: a community approach to sustainable agricultural development. CGIAR Research Program on Climate Change, Agriculture and Food Security (CCAFS), Copenhagen, Denmark. Available at http://ccafs.cgiar.org/publications/climate-smart-villages-community-approach-sustainable-agricultural-development#.VPVVX-FQ3o0. Accessed on 3 Mar 2015

Amarasinghe UA, Mutuwatte L, Surinaidu L, Anand S, Jain SK (2015) Reviving the Ganges water machine: why? Hydrol Earth Syst Sci Discuss 12:8727–8759

Amarnath G (2014) Earth observation data: monitoring floods and drought. Geospatial Today 6:15–19

Amarnath G, Ameer M, Aggarwal P, Smakhtin V (2012) Detecting spatio-temporal changes in the extent of seasonal and annual flooding in South Asia using multi-resolution satellite data. In: Civco DL, Ehlers M, Habib S, Maltese A, Messinger D, Michel U, Nikolakopoulos KG, Schulz K (eds) Earth resources and environmental remote sensing/GIS applications III: proceedings of the international society for optics and photonics (SPIE), vol 8538, Amsterdam, Netherlands, 1–6 July 2012. International Society for Optics and Photonics (SPIE), Bellingham, WA, USA, p 11

Campbell BM, Thornton P, Zougmoré R, van Asten P, Lipper L (2014) Sustainable intensification: what is its role in climate smart agriculture? Curr Opin Environ Sustain 8:39–43

CGWB (2013) Master plan for artificial recharge to ground water in India. Central Ground Water Board (CGWB), Ministry of Water Resources, Government of India, p 225. Available at http://cgwb.gov.in/documents/MasterPlan-2013.pdf. Accessed on 26 July 2015

Choppra K (2005) The raining nature. Available at http://www.boloji.com/. Accessed on 26 July 2015

EM-DAT (2014) The international disaster database. Centre for Research on Epidemiology of Disasters. Université catholique de Louvain, Brussels, Belgium. Available at http://www.emdat.net. Accessed 18 Feb 2015
FAO (Food and Agriculture Organization of the United Nations) (2010) "Climate-smart" agriculture: policies, practices and financing for food security, adaptation and mitigation. Food and Agriculture Organization of the United Nations (FAO), Rome, Italy. Prepared as a technical input for The Hague conference on agriculture, food security and climate change held on 31 Octo–5 Nov 2010, p 48. Available at http://www.fao.org/fileadmin/user_upload/newsroom/docs/the-hague-conference-fao-paper.pdf. Accessed on 27 July 2015
FAO (Food and Agriculture Organization of the United Nations) (2015) FAO success stories on climate-smart agriculture: CSA on the ground. Food and Agriculture Organization of the United Nations (FAO), Rome, Italy, p 28. Available at http://www.fao.org/3/a-i3817e.pdf. Accessed on 27 July 2015
Gale I, Neumann I, Calow R, Moench M (2002) The effectiveness of artificial recharge of groundwater: a review. British Geological Survey Commissioned Report, CR/02/108N, p 59
GoI (Government of India) (2008) National action plan on climate change. Prime Minister's Council on Climate Change, Government of India, p 49
Gosain AK, Rao S (2007) Impact assessment of climate change on water resources of two river systems in India. Jalvighyan Sameeksha 22:1–20
IPCC (Intergovernmental Panel on Climate Change) (2001) The scientific basis. Contribution of working groups I to the third assessment report of the intergovernmental panel on climate change. Cambridge University Press, Cambridge, UK, p 881
IPCC (2014) Climate change 2014: mitigation of climate change. In: Edenhofer O, Pichs-Madruga R, Sokona Y, Farahani E, Kadner S, Seyboth K, Adler A, Baum I, Brunner S, Eickemeier P, Kriemann B, Savolainen J, Schlömer S, von Stechow C, Zwickel T, Minx JC (eds) Contribution of working group III to the fifth assessment report of the intergovernmental panel on climate change. Cambridge University Press, Cambridge, UK, and New York, USA
Jat RK, Sapkota TB, Singh RG, Jat ML, Kumar M, Gupta RK (2014) Seven years of conservation agriculture in a rice–wheat rotation of eastern Gangetic plains of South Asia: yield trends and economic profitability. Field Crops Res 164:199–210
Karthikeyan B, Pavelic P (in prep.) Identifying priority watersheds to mitigate floods and droughts by novel conjunctive water use management strategies. Submitted to the Environ Earth Sci J Revised manuscript under review
Kumar MD, Patel A, Ravindranath R, Singh OP (2008) Chasing a mirage: water harvesting and artificial recharge in naturally water-scarce regions. Econ Political Wkly 43(35):61–71
Lipper L, Thornton P, Campbell BM, Baedeker T, Braimoh A, Bwalya M, Caron P, Cattaneo A, Garrity D, Henry K, Hottle R, Jackson L, Jarvis A, Kossam F, Mann W, McCarthy N, Meybeck A, Neufeldt H, Remington T, Sen PT, Sessa R, Shula R, Tibu A, Torquebiau EF (2014) Climate-smart agriculture for food security. Nat Clim Change 4:1068–1072
McCartney M, Smakhtin V (2010) Water storage in an era of climate change: addressing the challenge of increasing rainfall variability. Blue paper. International Water Management Institute (IWMI), Colombo, Sri Lanka, p 14
Milly PCD, Betancourt J, Falkenmark M, Hirsch RM, Kundzewicz ZW, Lettenmaier DP, Stouffer RJ (2008) Stationarity is dead: wither water management? Science 319(5863):573–574
Neate P (2013) Climate-smart agriculture success stories from farming communities around the world. CGIAR Research Program on Climate Change, Agriculture and Food Security (CCAFS); Technical Centre for Agricultural and Rural Cooperation (CTA), Wageningen, Netherlands. Available at https://cgspace.cgiar.org/bitstream/handle/10568/34042/Climate_smart_farming_successesWEB.pdf. Accessed on 27 July 2015
Nelson GC, Robertson R, Msangi S, Zhu T, Liao X, Jawajar P (2009) Greenhouse gas mitigation issues for Indian agriculture. IFPRI discussion paper 00900. Environment and Production Technology Division, International Food Policy Research Institute (IFPRI), Washington, DC, USA

Pavelic P, Srisuk K, Saraphirom P, Nadee S, Pholkern K, Chusanathas S, Munyou S, Tangsutthinon T, Intarasut T, Smakhtin V (2012) Balancing-out floods and droughts: opportunities to utilize floodwater harvesting and groundwater storage for agricultural development in Thailand. J Hydrol 470–471:55–64

Prathapar SA, Sharma BR, Aggarwal PK (2012) Hydro, hydrogeological constraints to managed aquifer recharge in the Indo Gangetic plains. IWMI-Tata Water Policy Res Highlight 40:5

Reddy VR, Syme GJ (eds) (2015) Integrated assessment of scale impacts of watershed intervention: assessing hydrogeological and bio-physical influences on livelihoods. Elsevier, Amsterdam, Netherlands, p 432

Revelle R, Lakshminarayana V (1975) The Ganges water machine. Science 188(4188):611–616

Sakthivadivel R (2007) The groundwater recharge movement in India. In: Giordano M, Villholth KG (eds) The agricultural groundwater revolution: opportunities and threats to development (Comprehensive Assessment of Water Management in Agriculture Series 3). CABI, Wallingford, UK, pp 195–210

Sellars S, Nguyen P, Chu W, Gao X, Hsu K, Sorooshian S (2013) Computational earth science: big data transformed into insight. Eos Trans Am Geophys Union 94(32):277–278

Shah T (2008) India's groundwater irrigation economy: the challenge of balancing livelihoods and environment. In: Chopra K, Dayal V (eds) Handbook on environmental economics in India. Oxford University Press, New Delhi, India

Shah T (2009a) Climate change and groundwater: India's opportunities for mitigation and adaptation. Environ Res Lett 4(3):1–13

Shah T (2009b) Taming the anarchy: groundwater governance in South Asia. Washington, DC, USA. Resources for the future. International Water Management Institute (IWMI), Colombo, Sri Lanka, p 310

Shah T, Verma S, Durga N (2014) Karnataka's smart, new solar pump policy for irrigation. Econ Political Wkly 49(48):10–14

World Bank (2010) Deep wells and prudence: towards pragmatic action for addressing groundwater overexploitation in India. The World Bank, Washington, DC. Available at http://www-wds.worldbank.org/external/default/WDSContentServer/WDSP/IB/2010/03/04/000333037_20100304230610/Rendered/PDF/516760ESW0P0951round0Water129101110.pdf. Accessed on 30 July 2015

Downstream Implications of Climate Change in the Himalayas

David J. Molden, Arun B. Shrestha, Santosh Nepal and Walter W. Immerzeel

Abstract The Himalayas, Hindu Kush, Karakorum mountains and the Tibetan Plateau make up the Hindu Kush-Himalayan (HKH) region, an area that has more snow and ice resources than any other region outside of the Polar Regions (Fig. 1). The HKH region extends 3500 km over all or part of eight countries from Afghanistan in the west to Myanmar in the east. About 200 million people live in the HKH mountains, while 1.3 billion people depend directly or indirectly on waters that originate in the mountains in 10 major river basins. These mountains are under threat from climate change and other socio-economic changes that will pose a challenge for Asia's future. This chapter reviews the state of knowledge concerning the mountain's water resources, draws out implications for downstream users, and recommends key actions to be taken.

1 Introduction

The Himalayas, Hindu Kush, Karakorum mountains and the Tibetan Plateau make up the Hindu Kush-Himalayan (HKH) region, an area that has more snow and ice resources than any other region outside of the Polar Regions (Fig. 1). The HKH region extends 3500 km over all or part of eight countries from Afghanistan in the west to Myanmar in the east. The HKH region provides a range of ecosystem services that are important both for local communities, for Asia, and for the world. The HKH region is diverse in many dimensions, including numerous indigenous communities, high biodiversity and agro-diversity, and diversity in climatic regimes.

Water, energy and food security in Asia is highly dependent on mountain water supplies. About 200 million people in Asia live in the HKH Mountains, while 1.3 billion people live in the ten river basins—the Amu Darya, Brahmaputra, Ganges,

D.J. Molden (✉) · A.B. Shrestha · S. Nepal
International Centre for Integrated Mountain Development (ICIMOD), Kathmandu, Nepal
e-mail: david.molden@icimod.org

W.W. Immerzeel
FutureWater, Wageningen and Utrecht University, Utrecht, The Netherlands

A.K. Biswas and C. Tortajada (eds.), *Water Security, Climate Change and Sustainable Development*, Water Resources Development and Management, DOI 10.1007/978-981-287-976-9_5

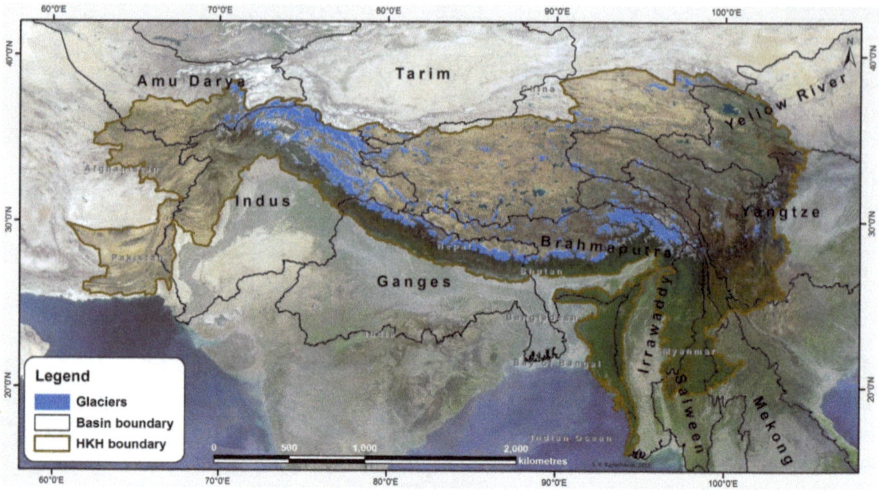

Fig. 1 The Hindu Kush-Himalayan (HKH) region and associated ten major river basins

Indus, Irrawaddy, Mekong, Salween, Tarim, Yangtze, and Yellow—which originates from the high-altitude mountains of the HKH region (Table 1, Eriksson et al. 2009) Natural and social activities and processes in upstream part of the basins affects the environment and influence the water availability for downstream users (Nepal et al. 2014a) who depend on mountain resources for growing food and obtaining energy, and are at risk of floods, increased sediment loads, and changes in water availability (Bruijnzeel and Bremmer 1989). Mountain communities, who are the custodians of these resources, are typically both economically and politically marginalized compared to people on the plains. Clearly, this relationship between upstream and downstream is crucial for better management of land and water resources at a basin scale. Moreover, many of the HKH Rivers cross international boundaries, providing an additional facet to the management of this upstream, downstream interface. The HKH region faces severe pressure from changes taking place across the world, such as climate change, air pollution, globalization and outmigration, all of which have put the adaptive capacity of people in mountain and plains to the test.

The HKH mountain resources are shared between many countries, and cooperation is critical for sustainable management of resources. The HKH countries include: Afghanistan, Bangladesh, Bhutan, China, India, Myanmar, Nepal, and Pakistan. Yet in spite of these pressures, there are many opportunities to adapt to change and build resilience, including sustainable use of resources, sharing of benefits of resource use between mountain and downstream communities and gaining income from high valued mountain niche products. There is a growing amount of scientific research and knowledge sharing that will help to realize these benefits and put them in practice.

Climate change is a matter of concern for the water resources and ice reserves of this region. One of the most pressing questions is how does climate change affect

Table 1 Characteristics of the river basins originating in the HKH region

River			River basin			
Name	Annual mean discharge (m^3/s)[a]	Glacier melt in river flow[b] (%)	Total area (10^6 km^2)	Population density (person/km^2)	Populatio (million)	Water availability (m^3/person/year)
Amu Darya	1376[a]	NA	0.65[c]	39	21	2081
Brahmaputra	21,261[a]	15.9	0.53[d]	182	119	5656
Ganges	12,037[a]	11.5	1.00[e]	401	407	932
Indus	5533[a]	41	1.11[f]	165	178	978
Irrawaddy	8024[a]	NA	0.43[g]	79	33	7742
Mekong	9001[a]	0.9	0.84[h]	71	57	4963
Salween	1494[a]	8.3	0.36[i]	22	6	7876
Tarim	1262[a]	$\leq$50	0.93[j]	7	8	4933
Yangtze	28,811[a]	~18	2.07[k]	214	369	2465
Yellow	1438[a]	~2	1.07[l]	156	147	308

Updated from Eriksson et al. (2009)

[a]The data were collected by the Global Runoff Data Centre (GRDC) from the following most downstream stations of the river basins: Chatly (Amu Darya), Bahadurabad (Brahmaputra), Farakka (Ganges), Pakse (Mekong), Datong (Yangtze), Huayuankou (Yellow)

[b]Estimation of the melt water contribution is difficult and varies in an upstream and downstream situation; approximates are given here. Values for Brahmaputra, Ganges, Indus, Mekong and Salween are based on Lutz et al. (2014)

[c]Tajikistan (72.8 %); Afghanistan (14.6 %); Uzbekistan (8.5 %); Kyrgyzstan and Turkmenistan (together 5.5 %)

[d]China (50.5 %); India (33.6 %); Bangladesh (8.1 %); Bhutan (7.8 %)

[e]India (79 %); Nepal (14 %); China (7 %)

[f]Pakistan (52 %); India (33 %); China (8 %); Afghanistan (7 %)

[g]Myanmar (91 %); China (5 %); India (4 %)

[h]China (21 %); Laos (25 %); Thailand (23 %); Cambodia (20 %); Vietnam (8 %); Myanmar (3 %)

[i]China (52 %); Myanmar (44 %); Thailand (4 %)

[j]China (97 %); Kyrgyzstan (2 %); Tajikistan (1 %); Pakistan and Afghanistan (together <1 %)

[k]China (100 %)

water availability, and what does this mean to downstream water users? The answer to this question will help to shape future water strategies. But the question also has to be answered in view of numerous changes taking place in the region. This chapter reviews current literature to provide new information to support future decisions and to identify critical knowledge gaps.

2 Water Resources in the HKH

The importance of water in the HKH and downstream is growing with increasing demands from growing population and economic development and at the same time an increasing number of floods and droughts. To meet future expectations for this water, it is vital to better understand the present and future availability of water.

Table 1 shows the key characteristic features of the ten big rivers originating from the HKH region. Both spatial and temporal variability of precipitation are high. About 70–80 % of the annual amount falling during the monsoon season in the central Himalayan whereas about 50 % in western part of the Himalaya (Nepal 2012; Rajbhandari et al. 2014). Heavy south-westerly monsoon rains often cause rivers to overflow their banks, sometimes annually, bringing devastation to the people living in the floodplain areas, while in the remaining months there are acute water shortages.

Water supply, use and scarcity varies considerably across the HKH region and associated plains. Although, the region seems abundant with water annually, at the temporal scale, there is a high variation between rainy and dry seasons. In mountains and hills, even with high rainfall, water shortage is a problem due to lack of storage options. For example, Cherrapunji in eastern Indian state of Meghalaya the location with the world's highest annual rainfall at an average of 11,987 mm (Murata et al. 2007), is sometimes referred to as a 'wet desert' because of the difficulty in accessing water in the dry season. Many of the river basins in the western Himalayas are physically water scarce (Molden 2007), with more than 75 % of river flows withdrawn for agriculture, industry and domestic purposes and supply barely able to meet demand. Some, like the Ganges, Brahmaputra, Salween and Irrawaddy basins, are considered 'economically water scarce', with less than 25 % of flow withdrawn for human purposes but with significant improvements in water infrastructure and management needed to make the existing resources available for use (Molden et al. 2007).

2.1 *Status of Cryosphere*

The Fourth Assessment Report by the Intergovernmental Panel on Climate Change's (IPCC 2007) showed that region was in data deficit, particularly related to cyrosphere data. The report included a controversial section on the Himalayan glaciers that raised alarm about the rate of melting and impact on rivers. Since then numerous studies have helped to fill some of the data gap. One such remote sensing based study revealed the extent of glacier coverage in the HKH region (Bajracharya et al. 2015), with a count of 54,000 glaciers and a total area greater than 60,000 km^2.

2.2 *Glacier and Snow Cover*

In Nepal and Bhutan remote sensing was used to monitor glacial changes over the last 30 years. This study indicates that more than 20 % decrease in glacier area over the time period. Newer glacier studies by Kääb et al. (2012; Bajracharya et al. 2014a, b) indicates widespread negative mass balance, with hot spots for glacier

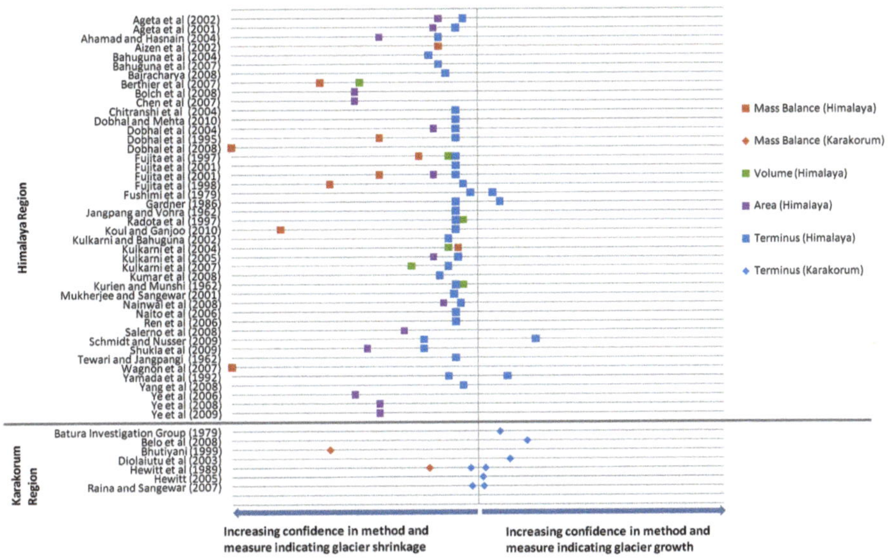

Fig. 2 Relative assessment of reviewer confidence in reported study outcomes according to method employed, clarity of reporting, and measurement type (*Source* Miller et al. 2011)

shrinkage are mostly in the eastern Himalayan areas. Miller and Rees (2011) compared recent studies on glaciers, together with the relative confidence levels for the measurement methods (Fig. 2). Their work shows that more glaciers are retreating than advancing. This information is still incomplete, with only a small number of glaciers monitored with ground data compared to the total of 54,000. Yet, not all glaciers are in retreat. Curiously, some glaciers are advancing, especially in the Karakorum region, a phenomenon described as the Karakorum anomaly (Hewitt 2005, 2011). The reason for this anomalous advance of glaciers is not yet understood. Most studies to date are based on terminus position and area change analysis, while mass change and volume change studies are limited. To complete the picture of the trends, status and future of glaciers, more objective and transparent analysis of glaciers are needed.

Large uncertainty remains regarding future glacier change. Lutz et al. (2014) provide a basin scale projection of glacier extent with a monthly time step from 2007 to 2050 for eight selected Global Circulation Models (GCMs). Strongest relative decreases in glacier area are expected for the Salween and Mekong basins (−33 to −56 % and −32 to −61 % for 4.5 and 8.5 RCPs respectively). The region has a smaller glacier area compared to the rest of the HKH. For the Indus basin, a decrease in glacier extent ranging from −20 to −28 % is projected based on the 8 selected GCMs. The Indus shows the smallest decrease, because it has the largest glacier area, although it has the highest projected increases in temperature and smallest increase (or decrease) in precipitation. Changes in glacier area in the Ganges and Brahmaputra show similar trends. The range of projections in these two basins is also similar as they have a similar range of changes in future temperature. Although the precipitation projections for the upper Ganges have a large spread, this does not result in a large spread in glacier area projections, showing that changes in temperature are the most important drivers of glacier area change.

Studies on the snow cover in the HKH region are more scarce than glaciers, although this has improved in recent years due to the advent of more space based sensors for snow cover monitoring. Gurung et al. (2011) estimated the average snow cover area in the HKH region for the period from 2000–2010 at 0.76 million km^2, which accounts for approximately 18.2 % of the total geographical area of the HKH region. The western part of the region has the most extensive snow-cover area on average because, in addition to having some of the highest elevations, it is at higher latitudes and is also more subject to the influence of winter westerlies (Bookhagen and Burbank 2010). In contrast to the noted reduction in the glaciers the long-term dynamics of snow cover in the HKH region are still not known, with different studies showing different results in both magnitude and direction (e.g. Gurung et al. 2011). It is clear that the historical data are still insufficient to generate statistically significant results.

Projections of future snow cover change in the HKH region is scarce. Böhner and Lehmkuhl (2005) used climate-model studies to predict that the snow cover of the Himalayan regions will decrease by 43 to 81 % by 2100 if the annual mean temperature at higher elevations in Asia increases by 1 to 6 °C as predicted by the IPCC. Similarly, the status and change of permafrost and its relative contribution to runoff are poorly understood, but could also be significant.

2.3 Current Hydrological Regime

The contribution of various input sources to the hydrology of the river basin has been a matter of growing scientific interest in the recent years and some fresh findings have provided important insights into the changing cryosphere and its impacts on water. The effects of changes in glacier melt, as summarized by Miller and Rees (2011), suggest that glacial melt is substantial, but that the glacial melt contribution to rivers varies from the dryer western part to the wetter eastern rivers (Fig. 3). Glacier melt contributes only 10 % of the annual river flow, on average, of the Ganges, and between 2 and 20 % for other basins of the Eastern HKH. Another study by Nepal et al. (2014b) estimated a 17 % contribution to flow in the Dudh Koshi catchment in eastern Nepal by using process oriented distributed J2000 model. The most comprehensive information on this to date is provided by Lutz et al. (2014) according to which glacier melt has the highest importance in the upper Indus basin (40 %), while the contribution of glacier melt to total flow is very small in the upper Salween (8 %; Fig. 3). The upper Ganges and upper Brahmaputra are dominated by rainfall-runoff, with glacier melt water contributing 16 and 11 % respectively. In the upper Salween and upper Mekong, rainfall runoff is also the most important, but the importance of snow melt is also substantial (27 and 35 %

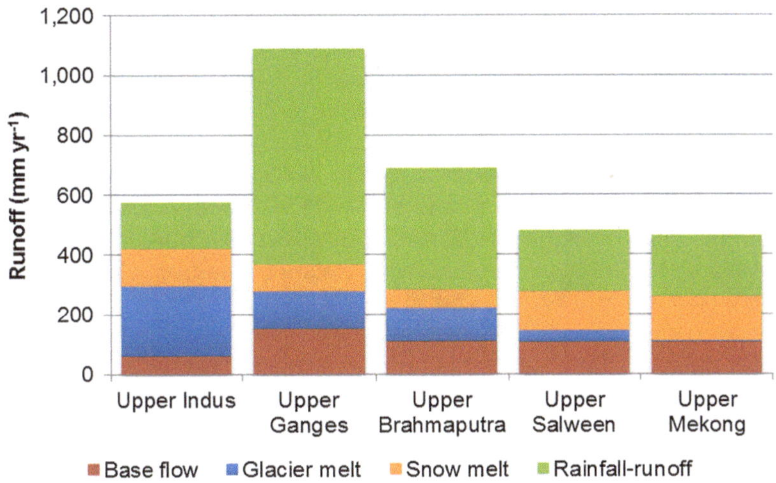

Fig. 3 Hydrological regime 1998–2007 in the upper river basins

respectively). The contribution of snow in the upper Indus is also high, around 22 %. Within Indus, sub-basin such as Kabul receives substantial amount of water from snow melt.

Rainfall and base flow contribute important fractions to the runoff even in the upstream catchments and they become even increasingly important in downstream portions of the basins. The groundwater contribution to the annual water budget of Himalayan river basins can be quite substantial, and for the eastern Himalayas it can be more significant than the contribution of glaciers (Andermann et al. 2012). Nepal et al. (2014b) estimated that the base flow contributes about 20 % to total annual discharge from the Dudh Koshi catchment in eastern Nepal based on hydrological modelling.

In summary, glacial melt contributes significantly to water discharge in areas of low rainfall such as the Indus. The Western Rivers of the HKH including the Indus, which do not receive much rainfall are the most vulnerable rivers to snow and glacier melt. The contribution of the glacier melt is the highest in the most upstream parts of the basins, near the glaciers.

3 Climate Change in the HKH Region and Its Impact on Water Resources

Observation records in the Hindu Kush-Himalayan (HKH) mountains are sparse often unreliable, and if available, often difficult to access. There are only around 100 regular stations collecting data above 3000 masl, and 40 above 4000 masl. Therefore climate change analysis in the region carries a high amount of uncertainty that could be reduced with more collection efforts and better data sharing. There are strong warming trends over the last decades across the HKH with warming rates higher than 0.01 °C/yr.

Unlike temperature, precipitation records do not show significant trends. Shrestha et al. (2000) found a cyclic pattern in the precipitation record in Nepal without any long-term trend. There is evidence that extreme precipitation events are increasing, for example in the central part of India (Goswami et al. 2006) and some indication of a similar change in parts of the HKH, but statistically robust trends have not been observed for the mountains. Immerzeel (2008) also analysed historic trends using global 100 year monthly high resolution data from 1900 to 2002, and concluded that the precipitation did not show any clear trend and was mainly determined by the monsoon.

Overall the projections of future precipitation suggest increase in precipitation in the future, particularly during monsoon. There is a high degree of uncertainty about future rainfall patterns, but interestingly, all climate scenarios are showing an increase in precipitation in the future.

Climate change is likely to increase the number of extreme events in the future. While comprehensive analysis of changes in the extreme events in the past and possible changes in the future is lacking, several studies indicate that such events are already increasing. Sivakumar and Stefanski (2011) suggested that South Asia has been experiencing more heatwaves, intense rains and floods, droughts and cyclones. Goswami et al. (2006) found significant trends in heavy and very heavy rainfall events in the central India. The Assessment Report 5 of IPCC suggests increase extreme events in the future in the region (IPCC 2013). Zhang et al. (2006) suggests that in future the occurrence of hot events is projected to be more frequent and the growing season will lengthen, while the occurrence of cold events is likely to be much rarer. There would be an overall increasing trend in extreme precipitation events in the middle and lower reaches of the Yangtze. The changes in the climate could mean more floods and droughts in the future (Lutz et al. 2014).

4 What Are the Impacts of Climate Change on Water Resources?

Changes due to climate change—in moisture inputs in the form of snow and rainfall, and temporary storage in the form of glaciers, ice, permafrost, snow, groundwater and wetlands—will have a profound effect on the flow regimes of the major rivers. While large uncertainties remain, we are beginning to get a picture of what will be the fate of HKH Rivers.

4.1 Water Availability

Lutz et al. (2014) used the future climate scenarios described in Sect. 3 to obtain future water availability scenario. In the upper Indus total runoff is likely to change by −5 to +12 % by 2050. The share of melt water stays the same, and the changes in total runoff are directly related to changes in precipitation. In the upper Ganges it is likely that total runoff increases by ~1–27 % by 2050. The share of melt water is decreasing and the share of rainfall-runoff in total runoff is increasing. Modelled stream flow in the upper Brahmaputra shows an increase by 2050 (0 to +13 %). The share of melt water is decreasing while the share of rainfall-runoff increases. In the upper Salween, it is likely that the total runoff changes by −3 to +19 % by 2050. The share of glacier melt and snow melt is likely to decrease, while the share rainfall-runoff increases. The runoff in the upper Mekong is predicted to increase by ~2 to 20 % by 2050. The share of melt water decreases while the share of rainfall-runoff increases. The trends are in general similar for RCP 4.5 and RCP 8.5 although the spread between the GCMs is large, especially for the Ganges, mainly due to the larger spread in precipitation projections in the GCMs used for RCP 8.5 compared to the GCMs used for RCP 4.5.

While the stable or increased water availability is envisaged at the large sub-basin or basin scale, at high elevation at the headwater of the rivers, where cryosphere is the predominant contributor to the flow, future water availability is likely to decrease. In the upper Indus Basin, for example, there are several examples of dysfunctional irrigation systems due to retreat and down wasting of glaciers and reduced flow.

Not all sub-basins in the upper catchments of the HKH are fed by cryosphere melt. The mid-hill river systems are predominantly rain fed and the main source of water for the communities in these catchments are groundwater fed springs. There are many examples of such springs drying out putting communities at water stress, but it is difficult to link this phenomenon with climate change. Both socio-economic and environmental changes are reported to be responsible for water stress situation.

4.2 *Glacial Lake Outburst Flood*

Glacial lakes often form between the frontal moraine and the retreating glacier or on the surface of the lower section of the glacier. These kinds of lakes are held back (dammed) by more or less unstable moraine complexes, and have a potential to breach their moraine dams. This phenomenon, in the Himalayas and elsewhere, has become known as glacial lake outburst flood (GLOF) and has the potential for generating extensive destruction in the valley downstream. The impact of such an outburst depends on the physical character of the dam, the lake size and depth and the rapidity of its drainage, and the nearby surroundings. Glacial hazards, such as ice avalanches, GLOFs, and debris flows, have caused severe damage in populated mountain regions in the HKH (and in many mountain areas throughout the world), and there is a concern that their frequency could increase as a result of accelerated glacial thinning and retreat. In certain circumstances, a GLOF can instantaneously release a huge amount of water and debris. This most likely would cause extensive effects on the downstream areas posing a threat to human lives and infrastructure.

Thus, GLOF risk assessment has become an issue of considerable significance that must be dealt with.

There are over 15,000 glacier lakes in the HKH, and over 200 are considered to be dangerous (Ives et al. 2010). Because of the changing situation, and because of the difficulty in getting accurate ground data, there is a need to continually update this information. There have been 34 recorded GLOF events in Bhutan, China and India region, while many more may have gone unrecorded (Ives et al. 2010). Among them 10 events were transboundary. According the glacial lake inventory updated in 2011 by ICIMOD (2011), there are 8790 glacial lakes occupying 801 km^2 area in Brahmaputra, Ganges and Indus river basins. 203 of these glacial lakes are potentially dangerous. With glaciers are projected to reduce by 20–55 % by 2050 is very likely that there will be increase in the number and areas of glacial lakes and there will be more GLOFs in the future. Glacial events in the past have caused severe damage in populated mountain regions in the HKH (and in many mountain areas throughout the world), and there is a concern that their frequency could increase as a result of accelerated glacial thinning and retreat. In certain circumstances, a GLOF can instantaneously release a huge amount of water and debris. This most likely would cause extensive effects on the downstream areas posing a threat to human lives and infrastructure. Communities close to the cryosphere can expect increased hazards including GLOFs. There are also hidden threats as water reservoirs within glaciers sometimes unexpectedly burst. Thus, GLOF risk assessment has become an issue of considerable significance that must be dealt with.

5 What Does It Mean for Water Supplies Downstream?

All across Asia, demand for water is growing for irrigation, cities and industries (Falkenmark 1997), and in many areas such as the Indus and Yellow river, growth in water use is already constrained by supplies. A key concern is about future supplies from mountain areas given the changes in the hydrologic cycle from climate change. Recent results from modelling exercises are beginning to shed light on the fate of future water supplies. The basic approach is to use the IPCC scenarios for future temperature and precipitation, downscale these to the HKH region, then use the data into hydrological models to understand future hydrology.

Immerzeel (2011) used fully distributed models, with input from a full range of global climate models. The model paid particular attention to the contribution from glaciers. Projections show a steady decline in glacial area, with an increase in temperature, and an increase in precipitation. The melting of glaciers combined with a decrease in glacial area leads to an increase in flows up to the period of 2050. However at a certain point of time in the future, as glacier mass declines, the glacier contribution to flow is likely to decrease (Rees and Collins 2006). This kind of trend has also been suggested in the Niyang River basin in south- east Qinghai (Zhang et al. 2011). Soncini et al. (2014) suggest that future flows are likely to increase until midcentury and then to decrease, but remaining mostly higher than control run values. Snowmelt is projected to occur earlier, while the ice melt component is expected to increase, with ice thinning considerably and even disappearing below 4000 m MSL until 2100.

The implication is that the overall annual flows are not likely to decrease with climate change in the future, and are in fact likely to increase somewhat. However, what is not certain is the extent to which the hydrograph will shift will also have significant impact on water management and agriculture. For example, changes in dry season flows could affect agriculture. Furthermore the degree to which extreme events will occur and will be translated into the hydrological regime can pose serious challenge to the water resources management. Other evidence (Goswami et al. 2006; Zhang et al. 2006) suggests that there will be greater variability in precipitation leading to more high flow events, and drought cycles. Moreover, glaciers are a form of natural storage that regulates flow and releases water in the dry season, and a loss in the storage could mean a loss in resilience.

6 What Are the Possible Impacts of Climate Change on Water Resources in Mountain Areas?

Many mountain communities are either directly dependent on glacier melt, or experiencing increased risk due to the hazards of glaciers (Fig. 4). Especially in the western part of the HKH, many mountain communities depend directly on melt runoff for irrigation and drinking water, building canals to tap into glaciers directly. In these situations, when glaciers shrink, the water source of the canals also dry. In the Gilgit region in Pakistan for example, one can observe a series of canals going higher and higher up mountain slopes where people have put effort to tap into glaciers at higher elevation as the glaciers have melted.

Another potential threat is the decreased slope stability caused by melting of permafrost. The frozen water of permafrost has an important function of holding soil and earth material together. However, when permafrost melts, there is an increased risk of landslides. Furthermore intense rains are also a trigger of landslides (Huggel 2009).

People living in the foothills of mountain areas have always had a struggle with water access as people live on the hillsides, and big rivers are in valley bottoms and difficult to tap. An important source of water for hill and mountain people are springs, proving supply for irrigation and household use. There is anecdotal evidence from across the HKH region that many of these spring sources are drying up. However, it is not clear whether this is due to climate change, increased use or environmental degradation (e.g. Tambe et al. 2012).

In 2014 in Nepal alone, there was an avalanche in the Everest Area causing the death of many climbers, a freak snowfall event that caused the death of several trekkers, there were major floods events in Western Nepal. A question is whether these events were caused by climate change. In fact for any individual event, there

Fig. 4 Irrigation canal drawing water directly from Passu glacier in Gilgit Baltistan Province of Pakistan

are many reasons for the loss of life and property that it is difficult to say that climate change is the culprit. Landslides, peak flows, glacier outburst floods, avalanches have always happened in mountains. The difference is that more people are trekking, climbing or living in flood plains and taking more risks.

7 What Are the Implications of Climate Change for Water Infrastructure Development?

The HKH region is already experiencing a high degree of disasters, and floods are the type of disaster that has caused the most death and displacement. In the future the flood hazard is likely to increase due to increased variability of rain, and increased threat from glacier lakes. Mirza (2002) projected a substantial increase in mean peak discharge in the Brahmaputra River, although by less than in the Ganges, based on climate change scenarios from four GCMs, which could lead to an increase in the occurrence of flooding of different magnitudes. However, it is important to recognize that disasters are caused when there is exposure to these hazards such as building houses and roads in flood plains. The implication is that we have to pay more attention to reduce the risk of disasters from floods and droughts in the future.

The demand for energy and hydropower is growing in the region, and in fact there has been a marked increase in the construction of hydropower facilities in mountain areas. A major question with climate change is whether our designs are adequate to take into consideration changes in hydrology or increased risk of high flows. Changes in glacier, snow and ice melt and increasing use of water need to be taken into consideration when planning hydropower development. More intense rainfall events in the future may lead to increased floods and sedimentation, leading to increased siltation for hydropower infrastructure to contend with. A modelling study of climate change impacts on the flow regimes of rivers in Bhutan concluded that the contribution to stream flow from glacier melt water will remain unchanged until about 2050 but will start to decline thereafter (Beldring and Voksø 2011). The impact of extreme events on the operation of hydropower infrastructure will be even more significant, as demonstrated by the drought of 2008 (Cheng et al. 2012), and uncertainty in design becomes more important. Improved watershed management, with compensation to watershed communities, is an important potential solution.

8 Economic and Social Implications

Perhaps the good news of the hydrologic forecasts is that there will be little change in the overall volume of water in the future, and there may be possible increases with higher rainfall. Of course there will variability within a year, and between years. If the water can be stored, and well managed, future water supplies from the mountains to downstream should not present any additional stress on urban and agriculture systems if these supplies are well managed, with water in during wet periods stored and put to use in dry periods. However, South Asian countries and India will still ace the huge management challenge in meeting additional demands for water for agricultural, urban and energy needs.

The biggest challenge will be the high variability in flows within the year, possible shifting monsoon patterns and patterns of river flow. For people downstream of mountains the implication is more floods and more droughts. These could have potentially severe impacts on agricultural systems, the environment, as well as urban supplies. This has repercussions on food supplies and food security. The poor are likely to be the most vulnerable to these extreme events, as we already witness today.

In many ways, we are already witnessing the impacts of mountain climate change. While it is not possible to pinpoint a particular flood or landslide to climate change, we see that these events are increasing, and likely to increase even further. Every year, there are severe floods in mountain areas, moving into downstream areas. The 2013 flood in Uttarakhand, the 2014 flood in Srinagar, and the surprising amount of rain in 2015 in Ladakh and Chitral and ensuing floods are some examples. These are devastating to local economies and livelihoods, and take considerable finances for responding and rebuilding. Moreover, maintaining social order and peace is more difficult in an environment where people are not thriving.

For mountain areas the situation is quite severe. Mountain people as mentioned before are susceptible to glacial melt, which can reduce water supplies for agriculture and city uses. Mountains, already a dangerous environment, are even more vulnerable to disasters. These are areas that are isolated and difficult to reach with relief, and materials for rebuilding. The poor, and women are the most vulnerable, especially in areas of high outmigration.

All countries of the South Asia and the Hindu-Kush Himalayas are experiencing rapidly increasing demand for energy, important for economic growth. Growth in hydropower facilities will be very important. However, these must be constructed taking into consideration the new hydrologic situation, plus the potential hazards of high flows, glacial lake outburst floods, and landslides. Existing hydropower plants have been damaged by these events, and designs and plans must better be able to take these into consideration. Shutdown of these facilities could be damaging to the economy in the short run.

It is possible to minimize the risk from climate change impacts on water in the mountains, and thus better support economic growth and societies. This will require better actions to manage future risks. This includes better plans and management that will take into account a range of hydrologic conditions. It will require more robust construction of flood control and hydropower infrastructure. It will require work with poor and vulnerable communities to improve their livelihoods and build resilience.

9 Action Needed

It is clear that climate change combined with changing mountain societies and environments presents new challenges. There are numerous actions that need to be taken now, both to adapt to changes, and to develop options to mitigate and reduce the impacts of climate change. Taking actions to mitigate global climate change to slow down the impacts of the mountains is on the top of the list.

It is very likely that we will face a future with increased environmental hazards that has repercussions to human safety and infrastructure development. Preparedness for disasters is a high priority. A key action is to develop "end to end" flood information systems to plan, prepare and respond to flood events (Shrestha 2008a, b; Shrestha et al. 2015). This includes risk mapping and zoning and being alert to new hazards as they arise. It includes gathering and sharing early warning data on rainfall and precipitation, and importantly sharing this quickly with people who will be affected; and then developing means to react when disaster happens. The HKH region can take inspiration from the cyclone early warning and response system in Bangladesh that has saved millions of lives (Paul 2009).

In line with disaster response, the key focus is on HKH communities who face high risk of floods. There is a large payoff in capacity building of local communities, developing communication systems, and to make sure that responses are available when disaster strikes. Women and children are at very high risk, especially now when there is so much outmigration. Women often do not have the skills to cope such as the basic skills of swimming or climbing trees. Community based early warning systems, operated and maintained by local communities show promise.

Changes in water regimes require special consideration for infrastructure development. Risks, including risks of GLOFs must be explicitly taken into account. Increased sedimentation from high rainfall events have to be taken into consideration, and watershed protection measures with compensation to watershed communities will become more imperative. The impact of extreme events on the operation of hydropower stations will be even more significant, as demonstrated by the drought of 2008 (Cheng et al. 2012), and uncertainty in design becomes more important.

Given the high stakes in terms of disaster risk reduction, and sustainable management of water resources, it is essential that our decisions are based on an adequate information base. This includes information about hydrologic components, and in addition, information about the environment and the social and economic situation of mountain people. There remains a dearth of data from high elevation, essential for a better understanding of changes. While there has been some progress, the region remains largely data deficit, and data sharing remains weak.

Most of the rivers emanating from the HKH region are transboundary. To adequately deal with disasters, to better plan for development, and to better manage water resources, improved cooperation and data sharing is essential (Rasul 2014). The countries of the HKH region have had some success in sharing real-time hydrological data, primarily through bilateral agreements, and this has proved useful in flood forecasting. However, achievements have been limited with regard to the sharing of real-time data and information on a regional scale, so critical for flood management (Shrestha et al. 2013).

A vital message is to focus on mountain people, who ultimately are the caretaker of the source of water resources, and other vital resources downstream. Often mountains are viewed as the source of resources for people downstream, and mountain people are often marginalized in decision making. The result is that benefit flows and development, often bypasses mountain people. For a sustainable future and a healthy mountain environment, investing and sharing benefits with mountain people will have a huge payback.

References

Andermann C, Longuevergne L, Bonnet S, Crave A, Davy P, Gloaguen R (2012) Impact of transient groundwater storage on the discharge of Himalayan Rivers. Nat Geosci 5:127–132

Bajracharya SR, Maharjan SB, Shrestha F, Bajracharya OR, Baidya S (2014a) Glacier status in Nepal and decadal change from 1980 to 2010 based on Landsat data: ICIMOD, Kathmandu, Nepal

Bajracharya SR, Maharjan SB, Shrestha F (2014b) The status and decadal change of glaciers in Bhutan from the 1980s to 2010 based on satellite data. Ann Glaciol 55(66):159–166

Bajracharya SR, Maharjan SB, Shrestha F, Guo W, Liu S, Immerzeel W, Shrestha B (2015) The glaciers of the Hindu Kush Himalayas: current status and observed changes from the 1980s to 2010. Int J Water Resour Dev 1–13

Beldring S, Voksø A (2011) In: Beldring S (ed) Climate change impacts on the flow regimes of rivers in Bhutan and possible consequences for hydropower development (Report No. 4). Norwegian Water Resources and Energy Directorate, Oslo

Böhner J, Lehmkuhl F (2005) Environmental change modelling for central and high Asia: Pleistocene, present and future scenarios. Boreas 34:220–231

Bookhagen B, Burbank DW (2010) Toward a complete Himalayan hydrological budget: Spatiotemporal distribution of snowmelt and rainfall and their impact on river discharge. J Geophys Res 115:25

Bruijnzeel LA, Bremmer CN (1989) Highland lowland interaction in the Ganges Brahmaputra River Basin—a review of published literature. International Centre for Integrated Mountain Development (ICIMOD), Kathmandu, Nepal

Cheng C-T, Shen J-J, Wu X-Y, Chau K (2012) Operation challenges for fast-growing China's hydropower systems and respondence to energy saving and emission reduction. Renew Sustain Energy Rev 16:2386–2393

Eriksson M, Xu J, Shrestha AB, Vaidya RA, Nepal S, Sandström S (2009) The changing Himalayas: Impact of climate change on water resources and livelihoods in the greater Himalayas. ICIMOD, Kathmandu

Falkenmark M (1997) Meeting water requirements of an expanding world population. Philos Trans R Soc B: Biol Sci 352(1356):929–936

Flügel W-A, Pechstedt J, Bongartz K, Bartosch A, Eriksson M, Clark M (2008) Analysis of climate change trend and possible impacts in the Upper Brahmaputra River Basin—the BRAHMATWINN Project. 13th IWRA World Water Congress 2008, Montpelier

Goswami BN, Venugopal V, Sengupta D, Madhusoodanan MS, Xavier PK (2006) Increasing trend of extreme rain events over India in a warming environment. Science 314:1442–1445. doi:10.1126/science.1132027

Gurung DR, Amarnath G, Khun SA, Shrestha B, Kulkarni AV (eds) (2011) Snow-cover mapping and monitoring in the Hindu-Kush Himalayas. ICIMOD, Kathmandu

Hewitt K (2005) The Karakoram anomaly? Glacier expansion and the 'elevation effect, Karakoram Himalaya. Mt Res Dev 25(4):332–340

Hewitt K (2011) Glacier change, concentration, and elevation effects in the Karakoram Himalaya, Upper Indus Basin. Mt Res Dev 31(3):188–200

Huggel C (2009) Recent extreme slope failures in glacial environments: effects of thermal perturbation. Quatern Sci Rev 28(11):1119–1130

ICIMOD (2011) Glacial lakes and glacial lake outburst floods in Nepal. ICIMOD, Kathmandu

Immerzeel W (2008) Historical trends and future predictions of climate variability in the Brahmaputra basin. Int J Climatol 28:243–254. doi:10.1002/joc.1528

Immerzeel W (2011) Climate change impacts on the water resources of the Hindu Kush Himalayas. Paper presented at the authors' workshop for the regional report on climate change in the Hindu Kush–Himalayas: the state of current knowledge of ICIMOD, Kathmandu, Nepal

IPCC (2007) Climate change 2007: The physical science basis. Contribution of Working Group I to the fourth assessment report of the intergovernmental panel on climate change. Cambridge University Press, Cambridge (Solomon S, Qin D, Manning M, Chen Z, Marquis M, Averyt KB, Tignor M, Miller HL (eds))

IPCC (2013) Climate change 2013: the physical science basis. Contribution of Working Group I to the fifth assessment report of the intergovernmental panel on climate change. Cambridge University Press, Cambridge (Stocker TF, Qin D, Plattner G-K, Tignor M, Allen SK, Boschung J, Nauels A, Xia Y, Bex V, Midgley PM (eds))

Ives JD, Shrestha RB, Mool PK (2010) Formation of glacial lakes in the Hindu Kush-Himalayas and GLOF risk assessment. ICIMOD, Kathmandu

Kääb A, Berthier E, Nuth C, Gardelle J, Arnaud Y (2012) Contrasting patterns of early twenty-first-century glacier mass change in the Himalayas. Nature 488:495–498

Lutz AF, Immerzeel WW, Shrestha AB, Bierkens MFP (2014) Consistent increase in High Asia's runoff due to increasing glacier melt and precipitation. Nat Clim Change. doi:10.1038/nclimate2237

Miller J, Rees G (2011) Water availability: River discharge and glacial hydrology. Paper presented at Authors' Workshop of ICIMOD for the regional report on climate change in the Hindu Kush Himalayas. The State of Current Knowledge, Kathmandu, Nepal

Mirza M (2002) Global warming and changes in the probability of occurrence of floods in Bangladesh and implications. Glob Environ Chang 12(2):127–138

Molden D, Oweis TY, Steduto P, Kijne JW, Hanjra MA, Bindraban PS, Bouman BAM, Cook S, Erenstein O, Farahani H, Hachum A, Hoogeveen J, Mahoo H, Nangia V, Peden D, Sikka A, Silva P, Turral H, Upadhyaya A, Zwart S (2007) Pathways for increasing agricultural water productivity. In: Molden D (ed) Comprehensive Assessment of Water Management in Agriculture, Water for Food, Water for Life: A Comprehensive Assessment of Water Management in Agriculture. Earthscan and Colombo, International Water Management Institute, London

Molden D (ed) (2007) Water for food, water for life: A Comprehensive assessment of water management in agriculture. Earthscan and Colombo, International Water Management Institute, London

Mool PK, Bajracharya SR, Joshi SP (2001) Inventory of glaciers, glacial lakes, and glacial lake outburst flood monitoring and early warning system in the Hindu Kush-Himalayan Region. Nepal, ICIMOD 364P

Murata F, Hayashi T, Matsumoto J, Asada H (2007) Rainfall on the Meghalaya plateau in northeastern India—one of the rainiest places in the world. Nat Hazards 42:391–399

Nepal S (2012) Evaluating upstream–downstream linkages of hydrological dynamics in the Himalayan region. PhD Thesis, Friedrich Schiller University of Jena

Nepal S, Flügel WA, Shrestha AB (2014a) Upstream-downstream linkages of hydrological processes in the Himalayan region. Ecol Proces 3(1):1–16

Nepal S, Krause P, Flügel W-A, Fink M, Fischer C (2014b) Understanding the hydrological system dynamics of a glaciated alpine catchment in the Himalayan region using the J2000 hydrological model. Hydrol Process 28:1329–1344. doi:10.1002/hyp.9627

Paul BK (2009) Why relatively fewer people died? The case of Bangladesh's Cyclone Sidr. Nat Hazards 50(2):289–304

Rajbhandari R, Shrestha AB, Kulkarni A, Patwardhan SK, Bajracharya SR (2014) Projected changes in climate over the Indus river basin using high resolution regional climate model (PRECIS). Clim Dyn 44:339–357. doi:10.1007/s00382-014-2183-8

Rasul G (2014) Why Eastern Himalayan countries should cooperate in transboundary water resource management. Water Policy 16(1):19–38

Rees HG, Collins DN (2006) Regional differences in response of flow in glacier-fed Himalayan Rivers to climatic warming. Hydrol Process 20:2157–2169

Science Reviews (2008), doi:10.1016/j.quascirev.2008.06.007

Shrestha MS (2008a) Impacts of floods in South Asia. J South Asia Disaster Stud 1:85–106

Shrestha AB (2008b) Resource manual on flash flood risk management. Module 2: non-structural Measures. vol (includes CD insert), ICIMOD, 91 pp

Shrestha AB, Wake CP, Mayewski PA, Dibb JE (2000) Precipitation fluctuations in the Himalaya and its vicinity: An analysis based on temperature records from Nepal. Int J Clim 20:317–327

Shrestha AB, Eriksson M, Mool P, Ghimire P, Mishra B, Khanal NR (2010) Glacial lake outburst flood risk assessment of Sun Koshi basin, Nepal. Geomat Nat Hazards Risk 1(2):157–169

Shrestha A, Aryal R (2011) Climate Change in Nepal and its impact on Himalayan glaciers. Reg. Environ. Change 11(1):65–77

Shrestha AB, Wahid SM, Vaidya RA, Shrestha M, Molden DJ (2013) Regional water cooperation in the Hindu Kush Himalaya. Free flow: reaching water security through cooperation, In: Griffiths J, Lambert R (eds) UNESCO and Tudor Rose, pp 65–69

Shrestha MS, Grabs WE, Khadgi VR (2015) Establishment of a regional flood information system in the Hindu Kush Himalayas: challenges and opportunities. Int J Water Res Dev 1–15

Sivakumar MVK, Stefanski R (2011) Climate change in South Asia. climate change and food security in South Asia. Springer Science+Business Media B.V., Berlin, pp 13–30

Soncini A, Bocchiola D, Confortola G, Bianchi A, Rosso R, Mayer C, Lambrecht A, Palazzi E, Smiraglia C, Diolaiuti G (2014) Future hydrological regimes in the Upper Indus Basin: a case study from a high-altitude glacierized catchment. J Hydrometeorol 16:306–326

Tambe S, Kharel G, Arrawatia ML, Kulkarni H, Mahamuni K, Ganeriwala AK (2012) Reviving dying springs: climate change adaptation experiments from the Sikkim Himalaya. Mt Res Dev 32:62–72

Zhang M, Ren Q, Wei X, Wang J, Yang X, Jiang Z (2011) Climate change, glacier melting and stream flow in the Niyang River basin, southeast Tibet, China. Ecohydrology 4:288–298

Zhang Y, Xu Y, Dong W, Cao L, Sparrow M (2006) A future climate scenario of regional changes in extreme climate events over China using the PRECIS climate model. Geophys Res Lett 33: L24702

Economic Growth and Climate Proofing Asia Through Sustainable Water Resources Management

Bindu Lohani, Yasmin Siddiqi and Garrett Kilroy

Abstract Asia has demonstrated remarkable economic growth over the past decade. However, competing demands for finite water resources coupled with climate change impacts now pose a major threat. With India expected to have over half of its inhabitants residing in cities by 2050, achieving urban water security will become increasingly critical to enhance people's livelihood and to sustained economic growth. This chapter will consider innovative solutions to meet future demands for water focusing on future cities, whilst addressing climate change risk. Effective governance and policy reforms creating an enabling environment for integrated planning of urban and basin water resources are paramount.

1 Introduction

This chapter outlines the current state of water security challenges in Asia and India, with a focus on cities. It describes the current strategies in place to tackle these challenges, and provides a range of innovative examples of The Asian Development Bank (ADB) interventions in India and the region demonstrating how ADB investments can add value to these strategies to achieve better water security.

The 21st Century has been called the Asian century. This is clearly demonstrated with its rising share of the world's output at the same time lifting many millions out

This chapter has been prepared with the Support of G. Kim, Senior Director, Sustainable Development and Climate Change Department.

B. Lohani · Y. Siddiqi (✉) · Garrett Kilroy
Asian Development Bank, Mandaluyong, Philippines
e-mail: ysiddiqi@adb.org

B. Lohani
e-mail: bnlohani@adb.org

Garrett Kilroy
e-mail: gkilroy@adb.org

A.K. Biswas and C. Tortajada (eds.), *Water Security, Climate Change and Sustainable Development*, Water Resources Development and Management, DOI 10.1007/978-981-287-976-9_6

of poverty. If Asia maintains its current trajectory, it will double its share of global gross domestic product (GDP) to 52 % by 2050 (ADB 2011a).

Despite the rapid economic growth, over 75 % of the region is water insecure which if left unmanaged, pose a real threat to Asia's continued growth and prosperity (ADB 2013e). Unlocking the economy's growth potential will become increasingly reliant on water and other natural resources, which are fundamental to support industrial development, increasing populations and changing food demands as prosperity increases. Central to these challenges is water security. In Asia, water security is at risk from a range of interconnected pressures including population growth, urbanization, increasing water pollution, the over-abstraction of groundwater, water-related disasters, and climate change.

Interdependent with water security are the related challenges of food and energy security—the water-energy-food nexus. The competing demands for finite water resources are compounded by climate change. These increasing challenges threaten the livelihoods of billions of people. For example, energy price volatility contributes to food crises; expanding irrigation systems increases water and energy demands; and access to unreasonably cheap supplies of energy can lead to the depletion of water resources, further intensifying the impacts of droughts.

One of the striking characteristics of Asia's growth is that it is predominantly an urban phenomenon—half the world's urban population now lives in Asia and cities account for 80 % of Asia's GDP (UN HABITAT 2013). Today, seven out of the ten most populous cities of the world are in Asia. By 2050, Asia's urban population will nearly double from 1.6 billion to 3 billion (ADB 2011a). The case of India illustrates the centrality of the urban economy—which will provide 70 % of all new jobs and 70 % of GDP by 2030 (ADB 2013d). Delhi is projected to become by 2030 the second largest megacity in the world, after Tokyo, with a population of 36 million (United Nations, Department of Economic and Social Affairs, Population Division 2014).

The Census for India indicates that in 2011 the total population for India was 1.21 billion, with urban and rural proportions at 31.2 and 68.8 %, respectively (see http://censusindia.gov.in/2011census/censusinfodashboard/index.html). Some estimates indicate that the urban population of India will reach 55 % by 2050 (WEF 2011). Approximately 30 % of the population comprises internal migrants moving for a variety of reasons including marriage, employment and education (Census of India 2001). Provisional 2011 census data show that for the first time, India's urban population has grown faster than its rural population since the last census, with urban and rural decadal growth rates of 31.8 and 12.3 %, respectively. Figure 1 projects these trends to 2050.

Figure 1 highlights, as India's population grows, the convergence of urban and rural populations, with India's cities' proportion of the total population projected to exceed 50 % by 2050. Some estimates indicate this may be achieved as early as 2039 (Ministry of Urban Development and National Institute of Urban Affairs 2011). These new migrants to urban centers face numerous challenges including access to shelter, education, healthcare and basic amenities like water, sanitation and food (UNICEF 2012).

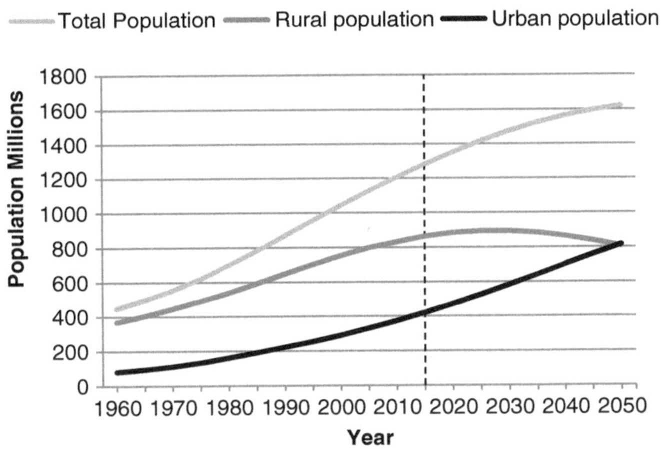

Fig. 1 Total rural and urban population growth for India. *Source* databank.worldbank.org

Table 1 Growth indicators for India's districts

Indicators	Rural	Transitional	Semi urban	Urban	India
Number of districts	153	219	84	33	640
Share of total GDP (% for 2012)	12	32	25	28	100
GDP growth rate (% for 2002–2012)	6.6	6.7	8.1	9.8	7.7

Source McKinsey & Company (2014b)

Recent analysis by McKinsey & Company assessed district, clusters and cities. It highlights the importance of urban districts in fuelling growth in India (McKinsey & Company 2014a, b). Table 1 lists a series of indicators developed from this analysis, which highlight the different stages of development of India's districts, with urban districts wealthier and growing at a faster pace.

The ADB Water Development Outlook (AWDO) for 2013 highlighted the under investment in growing Asian cities for public infrastructure and utilities, especially wastewater treatment as well as the over exploitation of water resources. Urban water security is one of the five AWDO dimensions, measuring how countries are creating better urban water services to develop vibrant, livable cities focusing on water supply, sanitation and drainage. The AWDO 2013 identifies South Asia as a hotspot where populations and economies are threatened by poor water security.

Figure 2 presents water security indices for India, People's Republic of China (PRC) and Indonesia, the three most populous countries in Asia. India's overall mean water security score is lower than both PRC and Indonesia. All three countries score low for urban water security, but India is particularly challenged where it scores only 1, the lowest, out of a 5-point index. Equally relevant here is India's low ranking (also 1) for household security, a dimension of water security that measures

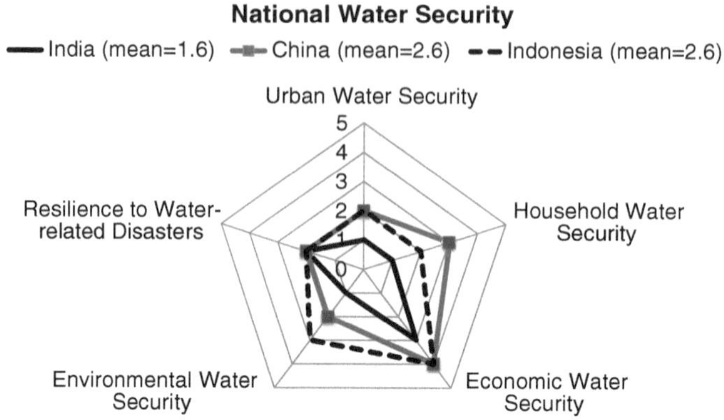

Fig. 2 Water security indices for India, PRC and Indonesia. *Source* Asian Water Development Outlook 2013

Table 2 Urban water and sanitation indicators for India, PRC and Indonesia

Indicators	India[a]	PRC[b]	Indonesia[c]
Nonrevenue water (%)	41	21	30
Share of energy cost as % of operational expenses	43	13	17
Water sold that is metered (%)	39	100	100
Duration of water supply (hours/day)	5.20	24.00	19.86

Source Danilenko et al. (2014)
[a]2009 data
[b]2012 data
[c]2004 data

the extent to which countries are satisfying their household water and sanitation needs and improving hygiene for public health in all communities. India's environmental water security, which is measured using a river basin health index, was also 1 and the Indian rivers studied had river health ratings amongst the poorest in the Asia and Pacific region.

Table 2 explores water security for these countries in a bit more detail, with a particular focus on urban water security. India has the highest level of nonrevenue water losses, which represents a loss not just of water but also of the energy used to produce and transmit that same water. This is also reflected in the share of energy cost as a proportion of operational expenses, which is much higher for India in comparison to PRC and Indonesia. Unlike India, full metering is installed in PRC and Indonesia, which offers a solid base for cost recovery and implementation of water efficiency measures. Overall, PRC has the most advanced indicators and this is also reflected in the level of service that can be achieved, such as delivering 24-hour water supply. As can be seen, India and Indonesia rank lower on household water security compared to PRC. Contrastingly, Indonesia ranks higher on

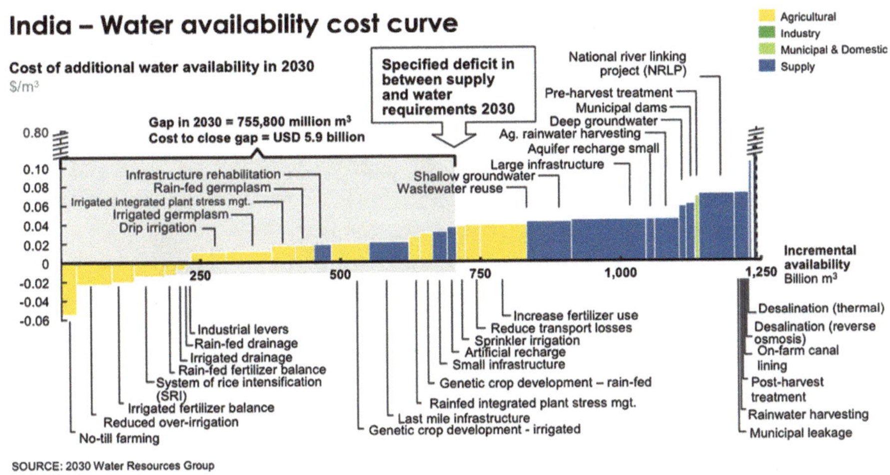

Fig. 3 India's water availability cost curve

environmental water security compared to PRC and India, suggesting much opportunity in the latter two countries to invest in water quality improvements.

Some of the key issues affecting water security in India include: uneven spatial and temporal distribution of rainfall, often resulting in floods in some areas and droughts in others; water use inefficiency and poor water productivity; unregulated and often unsustainable exploitation of groundwater resources; water pollution of surface waters due to point and non-point sources, and; inter-state disputes affecting transboundary water resources management (UNICEF, FAO and SaciWATERs 2013).

The Water Resources Group 2030 Report developed the water availability cost curve to explore options for closing the water supply-demand gap (Water Resources Group 2009). Figure 3 presents India's water availability cost curve. A key finding of this analysis is not only that India has options, as indicated by the range of measures presented, but also that much of India's water security can be achieved by improving the water efficiency and productivity of agriculture, which will comprise 80 % of demand by 2030. Some 80 % of the lowest cost measures needed to close the supply-demand gap can be achieved through agriculture. Overall if lowest cost measures are adopted, the overall annual expenditure in 2030 (including annualized capital and net operating expenditures) on managing India's water resources is US $5.9 billion.

PRC's water availability cost curve is presented in Fig. 4. In contrast to India, agriculture is estimated to comprise about 50 % of water demand in PRC by 2030 with industrial and urban water being the fastest growing sectors. PRC's lowest cost solutions lie in effectively managing industrial and urban water uses. Massive savings can be accrued wholly by employing industrial efficiency measures, most of which are negative costs and represent a net financial gain, with net annual savings of approximately US$21.7 billion. PRC's challenge will be getting the right mix of

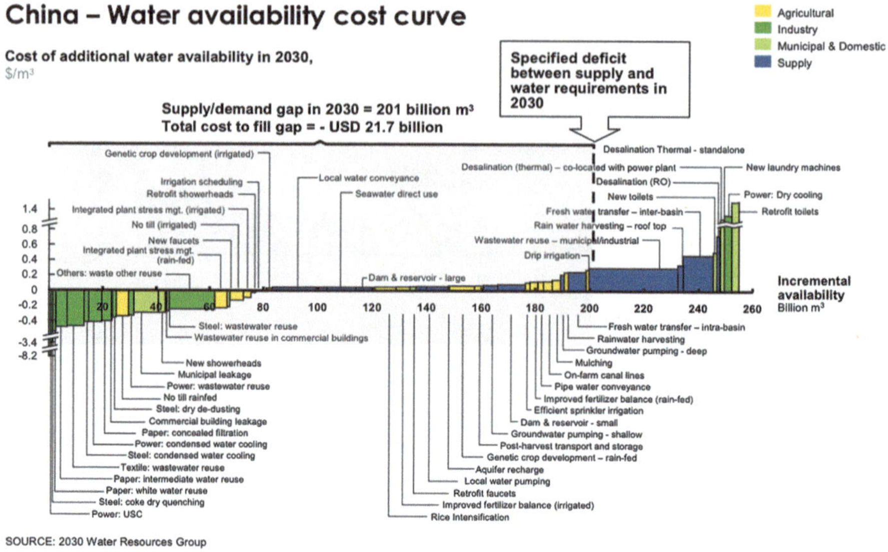

Fig. 4 PRC's water availability cost curve

measures, across this diverse country, that balance the competing demands of water and energy in the agricultural, supply, industrial, and municipal sectors.

With a decisive new leadership, which has already prioritized water on the development agenda, India is well poised to implement the reforms necessary to increase its economic growth in the coming years. Following the challenges outlined in the next section, this chapter discusses these reforms as they offer the opportunity to dovetail relevant investments, including ADB interventions. These interventions are designed to deliver urban water security and thereby contribute to securing the sustainability and resilience of Indian and Asian cities.

2 Challenges

2.1 *Water-Food-Energy Nexus*

Central to the challenges for Asian cities are the interconnected nexus issues of water, energy and food security, with the over-arching impacts of changes to climate and climate variability. Water security is typically acknowledged as the intrinsic element of this nexus (Beck and Walker 2013). This focus on water is not without sound justification—the synergies among the water-dependent sectors of agriculture, energy and water resources exemplify this interdependence.

The urbanization trends occurring globally, and particularly in India, provide a particularly pernicious backdrop to the nexus challenge. Increasing densely concentrated populations will drive demands for water, food and energy. Urban and peri-urban agriculture is expanding to meet the growing demands for food in cities, but the sector must compete with industry and municipalities for water and land resources as well as facing increasing urban pollution (IWMI 2014). Safe use of wastewater for irrigation and waste-to-energy initiatives are among the interconnected solutions required for tackling the nexus within this urban-rural interface.

2.2 *Water for Food Security*

Competing demands for water resources include increasing water requirements for grains to meet the demand for animal protein in diets as populations grow and prosper. Research indicates that India has the capacity to raise its agricultural yield growth from 2.0 %, its historical level, to 5.5 % annually over the next ten years using a variety of interventions including more efficient water management (McKinsey & Company 2014a). South Asian agriculture is extremely vulnerable to climate change with higher temperatures and changes in precipitation patterns posing a serious threat to food security (ADB 2014a). Demand management and increased water efficiency will be key requirements to address shortfalls in water supply. Vast growth in groundwater use in India has resulted in depleting aquifers in many areas often due to over-exploitation supported by energy subsidies.

Therefore in the case of India, the situation with regards to the nexus is well understood. The greatest gains are likely to be made in agriculture, where water demand is highest (about 80 % of total resource utilization) and irrigation efficiencies are low (typically 25–35 %). Surface water charges are also very low and unmonitored resulting in farmers having no direct economic incentive to be efficient or careful using water. Improved water use efficiency measures include participatory irrigation management, high efficiency irrigation systems, crop stress management and improved crop varieties. Conjunctive management is another opportunity for increasing irrigation efficiency and may be the best response to curb groundwater depletion, massive farm-power subsidies, droughts and poor water quality. Novel approaches like improving energy efficiency in irrigation, particularly pumping groundwater, are being implemented by ADB in India (see Sect. 5.6).

2.3 *Water for Energy Security*

Energy and water are linked in two ways: water is used in the production of nearly all types of energy, and energy is the dominant cost factor in the provision of water and wastewater services. Using wastewater (through anaerobic digestion) and solid waste (through waste to energy incinerators) for energy production require greater application in Asia. Energy for food production, particularly for pumping irrigation water, has had a significant impact on water and energy inefficiencies in Asia. Getting more crops per drop, using energy in food production more efficiently and adopting water efficient practices and technologies (including solar irrigation pumps) offer significant opportunities to revitalize agriculture in Asia to a more sustainable paradigm.

2.4 Demand Management for Water Security

The 2030 Water Resources Group (2009) estimate that with the total annual sustainable freshwater supply remaining static at 4200 billion cubic metres (m^3), the annual deficit for 2030 is forecasted to be 2765 billion m^3. This is 40 % of unconstrained demand. Assuming that present trends continue, India and PRC are forecasted to have shortfalls of 50 and 25 %, respectively. The challenge is how to meet these shortfalls. Overall, there is diminishing scope for increasing the supply side and the vast majority of water savings will have to be found on the demand side.

In India, agriculture dominates as the main user of water; however, industrial and domestic demands are increasing apace (Amarasinghe et al. 2007). Projected water demands in various sectors in India are presented in Table 3.

The Central Water Commission of India indicates 1869 billion cubic metres (BCM) of water resources are available, with 1121 BCM utilizable (690 BCM surface water and 431 BCM groundwater) (CWC 2013). India accounted for 17 % of the world's population in 2010 and an identical share of total water withdrawals. Between 1995 and 2010, per capita total water withdrawals increased to 32 m^3 and irrigation withdrawals increased to 25 m^3 (Amarasinghe and Smakhtin 2014).

Better demand management will become an essential foundation of any sustainable water resources management strategy. Vast growth in groundwater use in India has resulted in depleting aquifers in many areas often due to over-exploitation supported by energy subsidies. Surface water and groundwater should be managed conjunctively, as a single resource. Integral to achieving sustainable water demand management is getting the price right. Cost recovery approaches need the right balance to incentivize the adoption of efficient practices and technologies, while ensuring the sustainability of the investments.

Table 3 Projected water demand for various sectors in India (billion cubic metres)

Sector	2010	2025	2050
Irrigation	557	611	807
Drinking water	43	62	111
Industry	37	67	81
Energy	19	33	70
Others	54	70	111
Total	710	843	1180

Source National Commission for Integrated Water Resource Development (1999)

Table 4 Water indicators for urban and rural households in India and Gujarat

Water indicator	India			Gujarat State		
	Total	Rural	Urban	Total	Rural	Urban
Households with drinking water source within premises (% of households)	46.6	35.0	71.2	64.0	48.3	83.7
Households with latrine facility within premises (% of households)	46.9	30.7	81.4	57.3	33.0	87.7

Source Census of India 2011

2.5 *Improved Water Supply and Sanitation*

Progress has been made in a number of cities in providing inhabitants with clean, reliable, efficient and financially sustainable drinking water supply. For example, Phnom Penn Water Supply Authority has successfully reduced nonrevenue water losses from 60 to 6 % in 15 years through strong leadership and government support. ADB and others supported the rehabilitation of the distribution network. A move was also made to meter all connections and this was achieved by 2001 (ADB 2010). Similarly, Manila Water has succeeded in reducing water losses from 63 % in 1997 to 14 % in 2010, through privatization of the water industry, rehabilitating old pipelines and installing new ones—by 2011, Manila Water has replaced and laid a total of 3400 km of pipelines (ADB 2012).

Globally, some 2.5 billion people do not have access to improved sanitation and the world is unlikely to meet its Millennium Development Goals sanitation target of 75 %. In India, some 792 million people do not have access to an improved sanitation facility. However, progress is possible and the latest WHO/UNICEF report indicates that between 1990 and 2012, open defecation had decreased in South Asia from 65 to 38 % (WHO/UNICEF 2014). Nonetheless, most Asian cities do not have effective wastewater treatment systems. In the Philippines, for example, only 10 % of wastewater is treated while in Indonesia the figure is 14 %, in Viet Nam, 4 %, and in India, 9 % (ADB 2011b).

Table 4 outlines some of key water indicators for India and Gujarat. Urban households have greater provision of drinking water sources and latrine facility with premises. Whilst Gujarat demonstrates greater access to water supply and sanitation compared to national levels, there is significant scope for improvement across India.

2.6 *Urban Flooding and Water Quality*

According to the International Disaster Database (EM-DAT), floods over the last seventy years have shown the fastest rate of increase relative to other types of disasters. Asia reportedly registered about half of the total worldwide economic loss resulting from flooding. Floods appear to be occurring with increasing frequency and magnitude. Climate change impacts and sea level rise are expected to exacerbate these trends and low-lying coastal areas, where often assets and densely populated areas are concentrated, will be most vulnerable (Nicholls 2011).

2.7 *Financing Sustainable Urban Infrastructure*

Once key gaps in urban infrastructure are identified, a major hurdle for implementation by urban local bodies is credit worthiness and access to financing. For example, it is estimated that between 2012 and 2031, India will need to invest US $835 billion at 2009–10 prices, equivalent to 46 % of the country's 2012 GDP, to meet its urban infrastructure requirements (Chakrabarti 2014). While the entire amount cannot be financed through multilateral or bilateral agencies, urban local bodies have to access domestic capital markets to finance infrastructure improvement. This automatically implies improvement to municipal financial management. Such financial supports will become increasingly essential if the development of urban infrastructure is to keep pace with the demands arising from population growth and urbanization.

3 The Climate Imperative

Superimposed on all these challenges are the impacts from changes to climate and climate variability. Asia is on the frontline of climate change impacts both in terms of exposure and vulnerability. In parallel, climate impacts are most keenly felt on water resources. The latest IPCC report indicates that water scarcity is expected to be a major challenge for most of Asia due to increased water demand and poor management practices (Hijioka et al. 2014). The following key representative risks were identified for Asia during this latest assessment:

(i) Increased riverine, coastal and urban flooding leading to widespread damage to infrastructure, livelihoods and settlements in Asia;
(ii) Increased risk of heat-related mortality;
(iii) Increased risk of drought-related water and food shortage causing malnutrition.

The corollary of the concentration of economic growth in urban centers discussed earlier is that cities and their inhabitants generate 70 % of global energy use and energy-related greenhouse gas emissions. In addition, half of Asia's urban population lives in low-lying coastal zones and flood plains, areas which are most at risk from climate change impacts, including sea-level rise, storm surges and typhoons (Chakrabarti 2014; Hijioka et al. 2014). Therefore, Asian cities, as both a significant driver of global emissions and vulnerable zones for climate impacts, require a dual imperative of both mitigation and adaptation in their future development.

3.1 Economic Costs of Climate Change

The costs of inaction to adapt and mitigate against these impacts are enormous. The Stern Review in 2008 indicated that the costs to tackle global warming would equate to 2 % of global GDP, whereas the long-term damage of inaction would result in a 20 % reduction in global per capita consumption. The total costs of adaptation are significant but manageable. ADB's recent study for South Asia estimates that the total climate change cost, without global deviation from a fossil-fuel-intensive path, the region could lose an equivalent 1.8 % of its annual GDP by 2050, which will progressively increase to 8.8 % by 2100 on the average under the business-as-usual scenario (Ahmed and Suphachalasai 2014). If, however, global reductions in emissions manage to keep the global mean temperature rise below or within 2 °C, the region would only lose an average of 1.3 % of GDP by 2050 and roughly 2.5 % by 2100, highlighting the potential benefits of action. To avoid the damage and economic losses from climate change under the business as usual scenario, the region needs to provide an average adaptation expenditure of 0.48 % of GDP per annum ($40 billion) by 2050 and 0.86 % of GDP per annum ($73 billion) by 2100. On the lower global emissions trajectory, these costs reduce significantly to 0.36 and 0.48 % of GDP per annum, respectively, for 2050 and 2100.

3.2 Developing Urban Climate Resilience

The development pathway of the world's largest and fastest growing cities will be critical to the future of the global economy and climate. However, much of urban growth has been unplanned with significant economic, social and environmental costs. More compact and connected urban development, with reliable public transport systems, can create cities that are economically dynamic and more livable, and that have lower emissions. Such an approach to urbanization could reduce urban infrastructure capital requirements by more than US$3 trillion over the next 15 years (Calderon et al. 2014). The concept of urban climate resilience must be incorporated into urban development planning. Urban climate resilience embraces climate change adaptation, mitigation actions, and disaster risk reduction while recognizing the complexity of rapidly growing urban areas and the uncertainty associated with climate change (ADB 2014b). Such an approach is consistent with ADB's Green Cities Initiative, which is explored in more detail in the following section.

3.3 Climate Proofing

Effectively, what is required is the climate proofing of investments across sectors. Climate proofing is an approach to reduce the potential impacts of climate change through the identification of socio-economic and biophysical impacts and subsequent allocation of expenditure and project design for planned adaptation and risk management. ADB has developed a number of sectoral guidance documents on climate proofing investments, including for agriculture, transport and energy. These guidance documents provide a step-by-step methodological approach to assist project teams to assess and incorporate climate change adaptation measures into projects from these sectors.

Given the range and overlap of these challenges outlined above, strategic and integrated interventions are required. The next section focuses on interventions that tackle urban water security and sets out some of the strategic approaches being adopted in India and outlines some examples of good practice from the region.

4 Strategic Approaches in India

Whilst India's urban water security ranking is amongst the lowest in Asia, which threatens sustained economic growth, India is embarking on a number of strategies that will positively contribute to tackling these challenges. India's 12th five-year plan is focused on inclusive and sustainable growth and includes programs in drinking water and sanitation, provision of critical infrastructure in rural and urban areas as well as further implementation of the National Action Plan on Climate Change. Recent initiatives launched by the new Government of India build on these programs and are discussed below.

4.1 Swachh Bharat (Urban)

The *Swachh Bharat* Mission (SBM) represents a major campaign by the new government, a joint mission led by the Ministry of Urban Development (MOUD) and the Ministry of Drinking Water and Sanitation (MDWS), to clean 4041 statutory cities in India. The Rs. 62,009 crore (USD 10 billion) programme will be implemented over a five-year period, concluding notably in 2019—the year of the 150th anniversary of the birth of Mahatma Gandhi. The urban component of the SBM will be led by MOUD and will encompass support for the provision of household toilets, community toilets, public toilets and solid waste management, with significant private sector investment envisaged. As well as achieving behavioural change of citizens towards healthy sanitation practices, the mission also sets objectives to strengthen urban local bodies' implementation capacity and to foster an enabling environment for private sector investment. Other relevant urban strategies include plans to build on the Jawaharlal Nehru National Urban Renewal Mission, whose objectives include tackling urban poverty, improving service delivery and empowering urban local bodies, and *Rajiv Awas Yogana*, which aims to address urban slums.

4.2 Swachh Bharat (Rural)

In the rural context, SBM *Gramin* led by the Ministry of Drinking Water and Sanitation (MDWS) will build on India's Rural Sanitation and Hygiene Strategy 2010–2022, *Nirmal Bharat Abhiyan*, which set out a strategic vision to eliminate open defecation, adopt improved hygiene practices and safe disposal of solid and liquid wastes in rural India. A rural initiative, complimentary to the 'Smart Cities', has also been announced—*Saansad Adarsh Gram Yojana* or the Parliamentarian's Model Village Scheme, also known as SAGY—which aims to create 2500 smart villages by 2019 and will be led by the Ministry of Rural Development (MRD).

4.3 Smart Cities

In 2014, the Government of India has recently announced 'One Hundred Smart Cities' initiative, with allocation of Rs. 7060 crore (USD 1.1 billion) in the Union Budget. This programme represents a major opportunity to dovetail water-related strategies and improve urban water security. While the details are still being developed, it is likely that such an initiative will require to be underpinned by significant development in physical, social and institutional infrastructure. Relevant water-related elements of physical infrastructure will include strategies to address drinking water and sanitation provision deficits, waterway rehabilitation and water demand management.

4.4 Pooled Finance Model

In the context of urban development and renewal, the Government of India has significantly expanded the Pooled Municipal Debt Obligation Facility to Rs. 50,000 crore (USD 8 billion) to support the provision of infrastructure, including public transport, solid waste disposal, sewerage treatment and drinking water in the urban areas (Speech of Arun Jaitley 2014). This facility has enormous potential to empower urban local bodies to finance urban renewal works. The previous successful application of this pooled finance model for the Greater Bangalore Water and Sewerage project, which involved setting up Karnataka Water and Sanitation Pooled Fund, has demonstrated what can be achieved (Indo-USAID 2006).

4.5 Cleaning the Ganga

The new government has also re-energized the commitment to clean up the river Ganges and support the work of the National Ganga River Basin Authority and the National Mission for Clean Ganga. Urban and rural initiatives such as SBM and those outlined above will significantly contribute to reducing point source inputs to the Ganges and its tributaries.

4.6 Efficient Irrigation

A key tenet of India's agricultural development has been the Accelerated Irrigation Benefits Programme (AIBP). The AIBP was launched in 1996 to fast-track the implementation of ongoing major and medium irrigation projects which were in an advanced stage of completion. The Indian Government's 12th Five-Year Plan identifies both the importance of irrigation for agricultural growth and the potential available, and stepped up support for development of water resources (including on AIBP) to Rs. 1,09,552 crore (USD 17.6 billion). A key goal of the National Water Mission is to improve water use efficiency by 20 %—this represents a potentially huge water saving given that agriculture uses 80 %.

5 ADB Regional Experiences

The ADB currently invests about $2 billion per year in (urban and rural) water-related projects and about $4 billion in energy projects. In India, the resource envelope for sovereign operations is about $2 billion per year for the next 5 years (2013–2017) (ADB 2013a). The technical assistance programme will amount to about $8 million per year. India will remain ADB's largest single country borrower. ADB's proposed investment commitment to India includes: $551 million for agriculture and water resources management; $1.948 billion for energy; $1.125 billion for water supply and other municipal infrastructure and services. Some of the projects (either underway or forthcoming) funded under these investments, which tackle water security, are described in more detail below in the context of the challenges and India's strategic approaches outlined above.

5.1 Water, Sanitation and Urban Resilience

The Government of India's plan to develop smart cities offer the opportunity to dovetail water management strategies to simultaneously make them smart, sustainable and resilient. New ADB projects beginning in Karnataka (KIUWMIP and KISWRMIP) will demonstrate approaches to improve the water security of urban areas and the surrounding basins through institutional strengthening, improved water use efficiency, improved water resource planning, monitoring and service delivery. These measures can help reduce non-revenue water losses, improve drinking water supply and sanitation coverage, and achieve effective and equitable cost recovery for sustainable water resources management. A new project in Rajasthan is particularly well aligned with the Smart Cities Initiative and includes delivering key water sector reforms such as the provision of 24-hour water supply and rehabilitation and expansion of the sewerage network, while at the same time rationalizing water and sewerage tariffs for operational and maintenance cost recovery (Rajasthan urban sector development program). Resource and water use efficiencies embedded in the project include reducing non-revenue water and energy losses as well as recycling wastewater and energy generation from waste sludge.

Similar examples of ADB investments in integrated urban development projects can be found in PRC. The Xinjiang Integrated Urban Development Project includes the reuse of treated wastewater, which will diversify the city's water sources and allow the use of increasingly scarce potable water for only essential domestic purposes. The Yunnan Chuxiong Urban Environment Improvement Project combines both urban water infrastructure with flood risk resilience with a view to achieving competitive, green, and inclusive city development.

Waste to energy (WTE) can be a particularly attractive option to increase urban efficiency of available resources. In 2009, ADB provided a $200 million loan to PRC Everbright Int. Ltd. to build and operate a series of WTE plants with clean technologies in secondary cities in the PRC. These are model plants reducing municipal solid waste volumes by 90 % and eliminate methane gas emissions from the treatment process. They also use advanced flue gas emission control to meet the world's most stringent standards, such as the European Union II standard for dioxin. By substituting the incineration of municipal solid waste for fossil fuel combustion and avoiding methane, it reduces GHG emissions and contributes to climate change mitigation. The Suzhou project in Jiangsu Province, funded by ADB, is the largest WTE operation in PRC processing 2500 tonnes of waste a day and generating 200 million kilowatt-hours of on-grid electricity annually.

5.2 *Rehabilitating Waterways and River Basins*

ADB is investing in a number of projects in the region involved to support rehabilitation of vital watercourses—approaches that could be effectively applied in India. ADB is currently supporting a project for the environmental rehabilitation of Anhui Chao Lake, the 5th largest lake in PRC and a key water source for industry, agriculture, transport, tourism and recreation (PRC). Key approaches being adopted to clean the Anhui Chao Lake include: (i) increased municipal point source pollution control; (ii) enhanced nonpoint source pollution control; (iii) improved institutional capacity of the Chao Lake Management Authority; and (iv) strengthened project management capacity.

The Citarum River and its tributaries in Indonesia's West Java are a vitally important water supply for both the city of Bandung and the greater Jakarta region, home to 25 million people. ADB has committed to provide Indonesia with a $500 million, multiyear loan to finance a wide-ranging cleanup and rehabilitation plan for the Citarum River basin (Indonesia). Key approaches of this project include: (i) institutions and planning for Integrated Water Resources Management (IWRM); (ii) water resources development and management; (iii) water sharing; (iv) environmental protection; (v) disaster management; (vi) community empowerment; and (vii) data, information and decision support; and (viii) Programme Management.

5.3 Green Cities

The ADB's Urban Operational Plan for 2012–2020 is tackling Asia's urbanization challenges by supporting investments in sustainable urban development based on the "3E" strategies of economic competitiveness, environmental sustainability, and equity. ADB's Green Cities approach is a key mechanism to operationalize these 3E strategies and an operational framework to guide the process of implementation has been prepared (Sandhu et al. 2014). Good practices and demonstrated approaches, within the operational framework, will become available as this initiative gets implemented allowing scaling out to other countries. Each Green City will be unique, each with its own vision and accordingly tailored priorities. However, common goals will include achieving a city that is more livable by focusing on improving air, land and water through management of the environment, economic competitiveness and equity.

5.4 Flood Risk Management

ADB's investment in urban flood management from 1990 to date totals $1.7 billion, the majority of which are in Bangladesh and the PRC. Urban projects are increasingly covering flood management involving measures that are both structural (embankments, levees, river channelling, etc.) and non-structural (early warning systems, flood forecasting, flood preparedness plans, public awareness campaigns, etc.). ADB's recent publication "Unflooding Asia" offers examples of recent flooding in Manila, Bangkok and Beijing and offers solutions to manage floods in urban areas using the Green Cities Way, which promotes integrated planning of urban water, flooding, solid waste, storm water and waste water (Vojinovic and Huang 2014). Embedding multi-functionality in the design of green cities, e.g. so that a park may serve both a recreation and storm drainage function, will help ensure that future cities are more sustainable, efficient, adaptive and resilient.

5.5 Improved Irrigation Efficiency and Productivity

ADB is funding a $58 million project to realize the full production potential of large-scale irrigation schemes in Bangladesh (Amarasinghe and Smakhtin 2014). The project will improve water use efficiency and productivity through performance based management contracts, with aims for 100 % cost recovery mechanisms to ensure sustainability. In addition, solar irrigation pumps will be piloted to reduce energy demand.

Water availability remains one of the greatest challenges to agricultural development in the PRC. A recent project supported by ADB (Comprehensive Agricultural Development Project) has being tackling this water-food nexus issue by promoting participatory irrigation management by establishing and empowering water users associations. In addition, irrigation tariff reforms (moving from irrigated area based tariff to water volume based tariff) and water use efficiency technologies are also being implemented.

ADB is funding a number of projects in India (like the Orissa Integrated Irrigated Agriculture and Water Management Investment Program and the Karnataka Integrated and Sustainable Water Management Investment Program) and the region to improve water use efficiencies and make savings through technologies and improved management approaches, some of which are outlined in the following paragraphs.

5.6 Rationalizing Power Subsidies

ADB is funding a $200 million project in Punjab State to rationalize the agriculture power subsidies to incentivize more efficient power use (Punjab development finance program). Free power to farmers has led to an unsustainable development paradigm without matching growth in food-grain yields. These power subsidies have led to declining groundwater table, necessitating the use of higher capacity pump sets, which, in turn, consume more power and result in higher power subsidies. Reforms include segregating agricultural feeder lines and 100 % metering, which will incentivize judicious use of energy. By targeting the power subsidy for agriculture in an efficient manner, a major benefit of the programme will be greater conservation of energy and water.

5.7 Agriculture Value Chains

One of India's leading spice producers, Akay Flavours and Aromatics, will receive ADB assistance of $16.5 million to expand its business, invest in climate-resilient farming technologies, and open up income opportunities for thousands of poor farmers in India and Cambodia. In India, the project will help Akay finance working capital for a new steam sterilization plant, expand research and development of spice products with health benefits, and develop a processing plant and marketing and distribution network for seasoning products. In Cambodia, the project will support investment in organic spice farming using drip irrigation and rainwater harvesting on Akay's model farm and in a new processing plant.

5.8 Financing Infrastructure Investment

Public Private Partnerships (PPP) offer scope for implementing water and wastewater projects that have proved challenging under typical public models. ADB has been helping to improve the institutional framework for PPP to promote private investment in rural areas in India. The ADB-Government of India PPP Initiative, Mainstreaming PPPs in India, is run jointly by the government and ADB staff from the central PPP cell in the Department of Economic Affairs (see http://www.adb.org/countries/india/public-private-partnerships/implementation).

5.9 The Gujarat Approach

Gujarat has been a leading light in India in creating the enabling environment for PPP investments. The Economist Intelligence Unit national assessment report evaluating the environment for PPP in Asia and the Pacific included Gujarat State in recognition of the state's rich body of experience in implementing infrastructure PPPs (Economist Intelligence Unit 2011). ADB also has had a key role in supporting the rationalization of the state electricity grid, which, through the Jyotirgram Yojana scheme, resulted in separate feeders for both agriculture and industry/household purposes, and, which played an important part in reducing groundwater overexploitation in the state. ADB has also provided technical assistance in revising the Provision of Urban Amenities in Rural Areas Scheme so that rural infrastructure can be provided in a cost-effective manner through the PPP modality.

Overall, Gujarat rates highly amongst Indian states in creating a positive enabling environment for investment and improved governance systems. Its success in achieving this has been attributed to the following key principles (Kohli and Sood 2010), which offer scope for emulation across India and elsewhere: (i) robust leadership; (ii) empowering civil servants to act as public servants; (iii) decentralization, such as the autonomy provided for municipal corporations; (iv) an open government, with the removal of bureaucratic barriers; (v) citizen-centred government services providing for ready access to services and public officials; (vi) a focus on results rather than launches, and; (vii) the active pursuit of private partnerships for major initiatives.

One such urban water project with a PPP approach was the Surat Wastewater Recycling Project, a joint initiative between the ADB, the Government of Gujarat and the Surat Municipal Corporation. The project focused on treating municipal wastewater to tertiary level to deliver water of sufficient quantity and quality for a nearby textile industry. Thus, industrial water security was delivered without use of precious potable water. The attraction of using a PPP approach here was to balance risk appetite of international and national partners without comprising project quality and performance obligations. In this particular case, the PPP approach was exemplary in its design and bidding process, but ultimately faltered due to a lack of political support. Nonetheless, it demonstrated how the private sector could be actively engaged through a transparent project development and bid process to set up a financially viable, environmentally desirable and more reliable structure for service delivery.

A recent ADB project in Gujarat has demonstrated what can be achieved through private sector partnerships. Since 2004, ADB has partnered with the Gujarat government and the private sector to establish a solar power park in Charanka, a remote village in an area with huge solar potential—an environmentally sustainable solution to India's power challenges. This solar park is set to be the world's largest photovoltaic power station, with a capacity of 500 megawatts on its completion by the end of 2014. The independent power producers handled the design of the park, ADB shared its expertise in innovative financial solutions, and the national and Gujarat government mitigated risks through favourable policies and tariffs, land issue settlement, and park promotion (ADB 2013b). The project is a sound demonstration of what proper planning and cooperation among local communities and the public and private sectors could do to transform a remote desert village into a thriving locality.

Available funds in ADB for tackling water security and nexus issues include: Urban Climate Change Resilience Trust Fund, Water Financing Partnership Facility, Urban Financing Partnership Facility, and Climate Investment Funds.

6 Opportunities

Cities are the engines of Asia's future economic growth. Urban water security for Asian cities needs urgent attention—the economic costs for inaction will be immense. Bold, immediate and decisive actions are required for demand management through policy reforms and embedding integrated water resources management institutionally.

6.1 Good Governance

Governance systems must be put in place to create the enabling environment for reforms to foster this growth and to facilitate integrated planning across water-dependent sectors. Good governance systems will be characterized by robust and consistent leadership over the long term. Similarly, interventions should be coupled with early high-level cross-sector policy engagement and institutional capacity building so that authorities are suitably empowered for effective governance.

6.2 Water-Energy-Food Nexus and Climate Change

There must be integrated approaches to managing water resources in the region across urban, peri-urban and rural scales. Managing water across the water-energy-food nexus, together with climate change, must be new paradigm. The nexus cannot be tackled without a holistic and encompassing process for managing water resources. For example, the Karnataka Integrated and Sustainable Water Resources Management Investment Program, India will support the government's IWRM road map by introducing measurement and water accounting to support planning water demand and utilization by various sectors. The concept of urban climate resilience must also be incorporated into urban development planning such as proposed with ADB's Green Cities Initiative.

6.3 Cost Recovery

The price must be right to recover costs and ensure sustainability. Progressive water tariffs such as those applied in ADB's Lao PDR water supply and sanitation project for northern and central regions show how services can be successfully extended to the poor (ADB 2013c). Institutions, sectors and citizens must be motivated through appropriately designed incentives, such as performance based contracts, that encourage the adoption of technologies and approaches as well as promote water conservation and wastewater reuse.

6.4 Learning Lessons

We must catalyze success by adopting good practices and novel solutions demonstrated in the region. For example, the impressive reductions in water losses achieved by the Phnom Penn Water Supply Authority and Manila Water demonstrate what can be done. A survey of 40 water utilities in Asia and the Pacific indicated the energy costs ranged from \$0.002 to \$0.12 cubic meter—reducing water losses, therefore, constitute huge potential for water saving and reduced energy costs (ADB 2012). Water security is also improved through maintaining reservoirs full and ready to deal with perturbations in water availability. Similarly, Singapore leads the way in demonstrating novel approaches to managing water, energy and waste in an integrated manner. A fundamental step in Singapore's success was having a water policy that integrates water supply and sanitation under one responsible agency (McIntosh 2014) Singapore's novel approach to reclaiming wastewater through reverse osmosis (NEWater) has effectively resulted in the achievement of a closed water loop. There is no reason why this approach cannot be replicated within the appropriate enabling environment and robust policy framework.

6.5 Innovation

The examples provided in this chapter highlight innovative approaches supported by ADB that can be effectively applied to tackle the range of nexus challenges threatening water security of cities in Asia and India. The key thematic competencies demonstrated in this chapter are urban water supply and sanitation, water resources management and rehabilitation, and urban climate resilience. The emphasis has to be on integration across these thematic areas to ensure that disparate water security threats are addressed in a holistic manner. The strategic programs developed in India are well placed to tackle these challenges. ADB can support the implementation of these strategies drawing on the experience, know-how and innovation from good practices across the region. In this way, a pathway towards clean, productive, efficient, equitable, resilient and livable cities can be achieved.

In conclusion, India and the region are poised to build on successive years of exceptional growth. However, the future is far from secure, particularly for its cities. Poor water, energy and food security coupled with climate risks all pose individual threats to growth. It is their interplay and interdependence, which poses the greatest challenge. It is through recognizing this interdependence, however, that real change can be enabled. Innovative solutions and technologies are available within the region. Developing the appropriate enabling environment, impetus and leadership to drive implementation is where the real challenge lies. In this way, cities and their surround basins can be developed that are water, energy and food secure, resilient, livable, competitive and equitable.

References

ADB (2010) Every drop counts. Asian Development Bank, Manila
ADB (2011a) Asia 2050: realizing the Asian century. Asian Development Bank, Manila
ADB (2011b) Fast facts: urbanization in Asia, Asian Development Bank, Manila
ADB (2012) Green cities. Asian Development Bank, Urban Development Series, Manila
ADB (2013a) Country partnership strategy: 2013–2017. Asian Development Bank, Manila
ADB (2013b) Thorough planning of solar park mitigates risk for investors, locals. ADB Knowledge Showcases, Asian Development Bank, Manila

ADB (2013c) Together we deliver—10 stories from ADB-supported projects with clear development impact. Asian Development Bank, Manila
ADB (2013d) Urban operational plan 2012–2020. Asian Development Bank, Manila
ADB (2013e) Asia water development outlook. Asian Development Bank, Manila
ADB (2014a) Assessing the costs of climate change adaptation in South Asia. Asian Development Bank, Mandaluyong City, Philippines
ADB (2014b) Urban climate change resilience—a synopsis. Asian Development Bank, Manila
Ahmed M, Suphachalasai S (2014) Assessing the costs of climate change and adaptation in South Asia. Asian Development Bank, Manila
Amarasinghe UA, Shah T, Turral H, Anand BK (2007) India's water future to 2025–2050: Business-as-usual scenario and deviations. International Water Management Institute, Colombo, 47p (IWMI Research Report 123)
Amarasinghe UA, Smakhtin V (2014) Global water demand projections: past, present and future. International Water Management Institute (IWMI), Colombo. 32p. (IWMI Research Report 156), Bangladesh Irrigation Management Improvement Project
Beck MB, Walker V (2013) On water security, sustainability, and the water-food-energy-climate nexus. Front Environ Sci Eng 7(5):626–639
Calderon F, Stern N (2014) Better growth better climate: the new climate economy report—the synthesis report. The Global Commission on the Economy and the Climate. World Resources Institute, Washington, DC
Census of India (2001) http://censusindia.gov.in/(S(t4r2n545z30eolrb5kqgorjc))/Census_And_You/migrations.aspx
Chakrabarti M (2014) The role of municipal bond market for financing urban infrastructure in India: an empirical assessment. Abhinav Natl Monthly Refereed J Res Commer Manag 3 (4):48–62
CWC (2013) Water and related statistics. Central Water Commission, Government of India, December 2013
Danilenko A, van den Berg C, Macheve B, Moffitt LJ (2014) The IBNET water supply and sanitation blue book 2014. World Bank, Washington, DC
Economist Intelligence Unit (2011) Evaluating the environment for public-private partnerships in Asia-Pacific. The 2011 infrascope: findings and methodology
Hijioka Y, Lin E, Pereira JJ, Corlett RT, Cui X, Insarov GE, Lasco RD, Lindgren E, Surjan A (2014) Asia. In: Barros VR , Field CB, Dokken DJ et al (eds) Climate change 2014: impacts, adaptation, and vulnerability. Part B: regional aspects. Contribution of working group II to the fifth assessment report of the intergovernmental panel on climate change. Cambridge University Press, Cambridge
IWMI (2014) Strategy 2014–2018—Solutions for a water-secure world. International Water Management Institute, Colombo
Karnataka integrated urban water management investment program and Karnataka integrated and sustainable water resources management investment program
Kohli HS, Sood A (2010) India 2039: an affluent society in one generation. Sage Publications, Thousand Oaks
McIntosh AC (2014) Urban water supply and sanitation in Southeast Asia a guide to good practice. Asian Development Bank, Manila
McKinsey & Company (2014a) From poverty to empowerment: India's imperative for jobs, growth, and effective basic services. McKinsey Global Institute, February 2014
McKinsey & Company (2014b) India's economic geography in 2025: states, clusters and cities. Identifying the high potential markets of tomorrow. Insights India, http://mckinseyinsightindia.com
Ministry of Urban Development and National Institute of Urban Affairs (2011) India's urban demographic transition: the 2011 census results (provisional)
National Commission for Integrated Water Resource Development (1999) Integrated water resources development. A plan for action. Volume I. Ministry of Water Resources, New Delhi
Nicholls RJ (2011) Planning for the impacts of sea level rise. Oceanography 24(2):144–157

Indo-USAID (2006) Financial institutions reform and expansion project-debt market component, www.niua.org
Sandhu SC, Singru RN (2014) Enabling GrEEEn cities: an operational framework for integrated urban development in Southeast Asia. ADB Southeast Asia Working Paper Series. No. 9, November 2014
UNICEF (2012) Overview of internal migration in India, http://www.unicef.org/india/1_Overview
UNICEF, FAO and SaciWaters (2013) Water in India: situation and prospects, http://www.unicef.org/india/Final_Report.pdf
United Nations, Department of Economics and Social Affairs, Population Division (2014) World urbanization prospects: the 2014 revision, Highlights (ST/ESA/SER.A/352)
UN HABITAT (2013) State of the world's cities 2012/2013. United Nations Human Settlements Programme (UN-HABITAT), Nairobi
Vojinovic C, Huang J (2014) Unflooding Asia: the Green Cities Way. ADB and IWA Publishing, London
Water Resources Group (2009) Charting our water future: economic frameworks to inform decision-making. World Bank, Washington, DC
WEF (2011) Water security—the water-food-energy-climate nexus. The World Economic Forum Water Initiative, http://www3.weforum.org/docs/WEF_WI_WaterSecurity_WaterFoodEnergyClimateNexus_2011.pdf
WHO/UNICEF (2014) Progress on drinking water and sanitation, http://www.wssinfo.org/

Business Case for Water

Joppe Cramwinckel

Abstract As competing demands for clean, fresh water (agriculture, households, energy generation, industrial use, ecosystems) continue to rise, the effects of climate change further exacerbate the challenges associated with water quality and availability creating new risks for businesses, governments, communities and the environment. Business can contribute to the alleviation of the strain on water by becoming water stewards by adopting an economically, environmentally and socially responsible water strategy. This means adopting values and practices that aim to safeguard long-term availability of clean water for all stakeholders in a watershed, prompted by recognition of water as an externality with a potentially material business risk. Industrial water reuse and managing agriculture withdrawals are important solutions on this path. In doing so, business has to become more actively involved in water governance.

1 Introduction

Water delivers economic, environmental and social benefits. For the most part, water used for domestic, agricultural and industrial needs is freshwater from rivers, lakes and aquifers; in some cases these demands are also met by brackish or briny water, recycled wastewater or desalinated saltwater. Water is generally renewable, replenished naturally by the water cycle. Its availability, however, is cyclical and distributed unevenly, due to wide differences in climatic patterns, geography and human use. Indeed, the human footprint on the hydrological cycle is both apparent on a global scale and acute in many localities—exposing vulnerabilities to both ecosystems and society.

Current projections estimate that global water demand will increase by approximately 55 % between 2000 and 2050, from 3565 to 5467 km^3 a year.

J. Cramwinckel (✉)
Water, World Business Council for Sustainable Development (WBCSD),
Geneva, Switzerland
e-mail: cramwinckel@wbcsd.org

A.K. Biswas and C. Tortajada (eds.), *Water Security, Climate Change and Sustainable Development*, Water Resources Development and Management, DOI 10.1007/978-981-287-976-9_7

The increase will mainly come from manufacturing (+400 %), the electricity production process (+140 %) and domestic use (+130 %) (OECD 2012). This leaves little scope to meet the increasing demand for irrigation water while respecting the needs of ecosystems. Without improvements in how water is managed and used, the world could face a 40 % supply gap by 2030 (Water Resources Group 2009).

Water sustains life, and plays a uniquely important role in the functioning of the whole Earth system. Although water is plentiful, 97.5 % of all the water on earth is saltwater and only 2.5 % is freshwater, of which 70 % is in the form of ice and snow cover. Around 30 % of the world's freshwater is stored underground in the form of groundwater, which constitutes about 97 % of all the freshwater that is potentially available for human use. Just 0.3 % of the world's freshwater can be found in rivers and lakes (http://www.unwater.org/statistics_res.html).

The global water cycle both influences and is influenced by weather and climate patterns, land cover and biodiversity, and the cycling of the essential nutrients that underpin all the world's ecosystems. Because of this interdependence, widespread local and regional human-induced and natural changes to freshwater resources now add up to a global concern.

1.1 The Scale of Demand on Water and Resulting Stress on Resources

An estimated 25 % of the world's river basins run dry before reaching the oceans due to the use of freshwater resources in the basins (Molden et al. 2007). These river basins, and many others around the world, have already reached a point where local freshwater availability does not keep up with demand. The increasing demand has taken a toll not only on surface water availability, but also on groundwater. Groundwater supplies water to billions of people, plays a central part in irrigated agriculture and influences the health of many ecosystems. Major aquifers around the world are being depleted at unsustainable rates (Gleeson et al. 2012), which, combined with declining surface water availability, is leading to food shortages, business disruption, competition and conflict (Ceres 2009; WWF 2009).

In addition, the water needs of ecosystems are often poorly recognized or ignored in water allocation decisions, so many important underlying ecosystem services are threatened, increasing longer-term sustainability risks with social and economic consequences. Furthermore, 7 out of 10 people are expected to live in urban centers by 2050 (WHO) and an estimated 1.8 billion people still lack access to safe water and over 4.1 billion people to adequate sanitation (Onda et al. 2012; Baum et al. 2013). Significant investments and coordination will be required to ensure water of appropriate quality and quantity for all users (i.e., domestic, agricultural, municipal, industrial, etc.) in both densely populated urban areas and rural communities.

Overall, the growing demand for water is estimated to leave over 3.9 billion (roughly 40 %) of the global population living in river basins under severe water stress by 2050 (OECD 2012). This statistics is alarming seeing the need to ensure

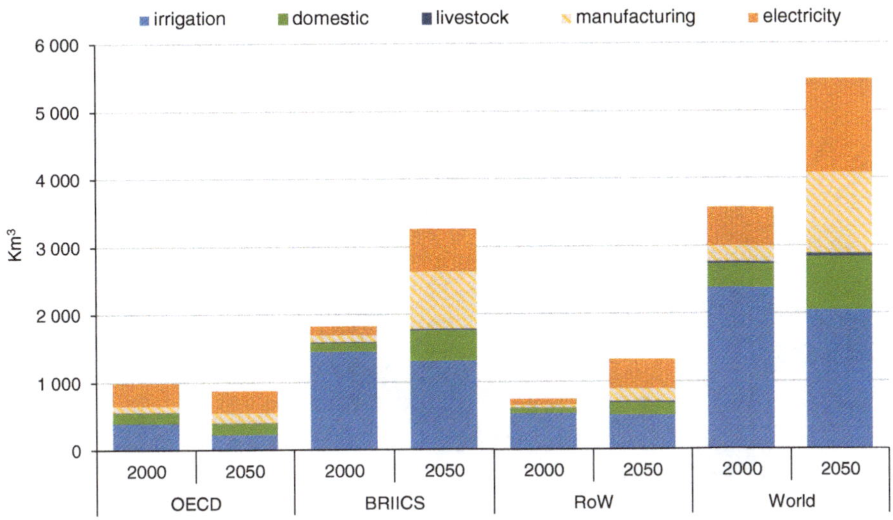

Fig. 1 Global water demand: baseline, 2000 and 2050. *Notes* This graph only measures *blue water* demand (see Box 5.1 of the *Outlook* report) and does not consider rain-fed agriculture. *RoW* rest of the world. *Source* OECD Environmental Outlook to 2050, The Consequence of Inaction

the distinct human right and basic human need for access to safe water and sanitation are distinct human rights and basic human needs, vital for life (Fig. 1).

The increasing and competing demands for water use—for example, from agriculture, households, energy generation and industrial use, as well as from the ecosystems that provide direct value to societies create new risks to businesses, governments, communities and the environment. These risks are broad-ranging, from conflicts, natural disaster and food security to financial, reputational, regulatory and operational (WWF 2009).

Water governance systems that allow for fair, equitable and sustainable allocation of water for all the stakeholders are a key element of sustainable water management. Such water governance should include inclusive stakeholder involvement, a watershed approach to the evaluation of water risks and challenges in a given watershed, and, for business, thorough evaluation of their short- and long term water risks both within their own operations, the environment in which they operate and in their supply chain (Fig. 2).

1.2 Key Interdependencies

Healthy and resilient ecosystems (e.g., wetlands, forests) play an essential role in the continued availability and quality of freshwater. For example, watersheds not only provide critical provisioning services, but also essential regulatory functions that can help reduce the costs of managing, treating and moving water. While

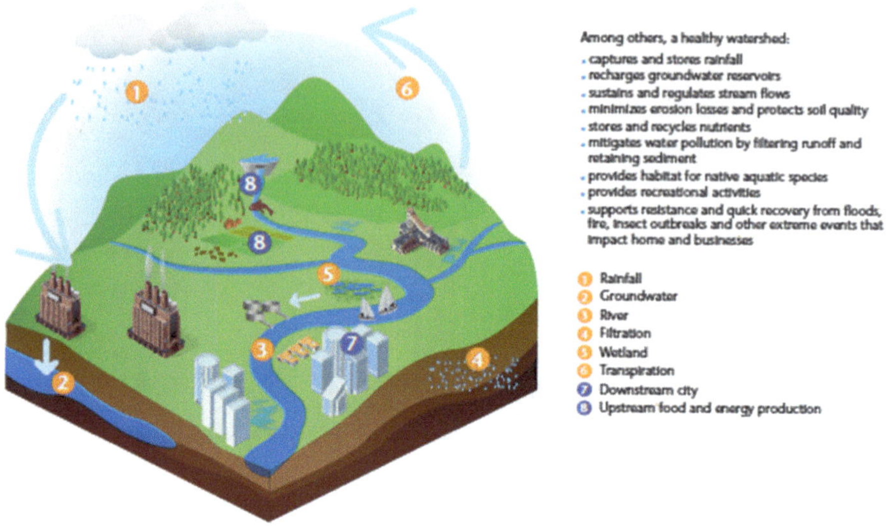

Fig. 2 Healthy watershed

businesses depend, to varying degrees, on the range of services a river basin and aquifer provide, they also have an impact on these services.

Water has direct impacts on the presence and distribution of nutrient elements in the environment. Eighty percent of municipal and industrial wastewater is currently discharged without treatment into the environment (UNEP and UN-Habitat 2010). This water may contain residues of fertilizers, limiting its use for domestic and industrial use. Persistent organic compounds and oxygen depleting nutrients seriously affect aquatic life.

Water is directly impacted by climate change, which is expected to exacerbate stresses on water resources. The widespread mass losses of glaciers and reductions in snow cover experienced over recent decades are projected to accelerate throughout the 21st century, potentially reducing water availability and hydropower potential. Changes in precipitation and temperature lead to changes in runoff and water availability. There is also high confidence that many semi-arid areas will suffer a decrease in water resources and drought-affected areas are projected to increase in extent. Regionally, demand for irrigation water is projected to rise as a result of climate change. Research suggests a significant future increase in the number of heavy rainfall events in many regions, including in some regions where mean rainfall is projected to decrease. The resulting increased flood risk poses challenges to society, physical infrastructure and water quality. Increased temperatures will further affect the physical, chemical and biological properties of freshwater lakes and rivers, with adverse impacts on water quality. In coastal areas, sea level rise will exacerbate water resource constraints due to increased salinization of groundwater supplies (IPCC synthesis report 2007).

1.3 The Crucial Water-Energy-Climate Change Risk

The earth system is projected to face multiple effects from gradual changes in the temperature and availability of water, as well as from extreme events tied to water, such as hurricanes and droughts.

These various 'water stresses' or events can affect the energy system in various ways.

For example, a higher temperature of cooling water can have a dual impact on electricity supply: through a direct reduction in thermal plant efficiency and through regulatory constraints limiting the temperature of water discharged by power plants. In addition, the availability of cooling water may also be reduced. These events would lead to reduced output by coal, gas and nuclear plants, as well as concentrated solar power plants. This constrained output may coincide with increased demand for power from increased needs for air-conditioning, water pumping for agricultural irrigation, etc. Hydro-reservoirs will also be affected, albeit with great variations from one location to the next. More significant snowfall and/or rain could enhance supply in some regions, while an increase in the number and severity of droughts would reduce it in others. Changes from snowfall to rainfall in winter could change the seasonal distribution of water availability.

Water availability can also affect fossil fuel supplies—some restrictions on the use of water for fracking have been applied in drought-hit areas in the US. Coal, oil and gas supplies are also exposed to weather-related supply disruptions, with Hurricane Katrina providing the most striking example in recent years. Even when the destruction of infrastructure is avoided, precautionary measures in the expectation of these events usually mean reduced output. Flooding has also affected mining operations or disrupted the transport of coal and oil—which is also exposed to low river water depth in times of droughts (Fig. 3).

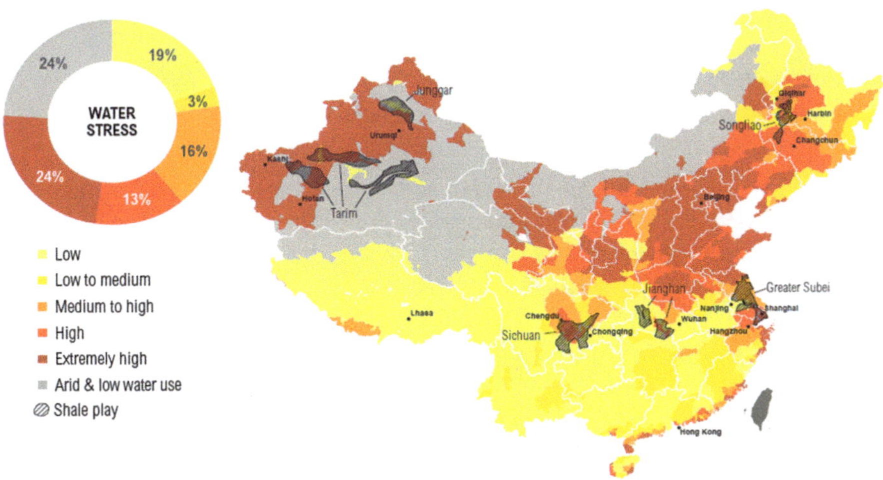

Fig. 3 China shale plays often overlap with areas of high water stress

Energy assets in coastal areas could also be exposed to more numerous/intense extreme weather events as well as a gradual sea level rise threatening both coastal power generation and transmission infrastructure. Some energy supply infrastructure is located in permafrost areas which would be affected by increased temperature, limiting road transport and access to production sites, as well as affecting the bearing capacity of the soil. This could disturb oil and gas extraction as well as pipeline infrastructure. Pipelines may also be vulnerable to weather-related accidents such as landslides and flooding.

There is therefore an intrinsic link between the challenge we face to ensure water security and other global issues, most notably climate change and the need to sustainably manage the world's rapidly growing demand for energy and food, as well as ecosystems. We cannot rely anymore about past trends, we need to understand what these changes mean in terms of ensuring adequate water supply in the future.

We need to develop strategies to cope with gradual long term climate change-induced impacts and uncertainties, as well as options for policy responses, and change the way they design and manage infrastructure in response to the risks of longer-term climate impacts on supply due to greater water scarcity; changes in average winter and summer temperatures; and sea level rise.

Huge efforts have led to improved seasonal forecasting but the predictability of the climate season one season ahead remains moderate, especially in mid-latitudes, including the North Atlantic and Europe. Recent studies have shown some potential improvements in predictability, in particular with variables linked to the water cycle, such as river flow and soil humidity.

Water impacts will vary regionally, thus particular impact will need to be modelled at the local level. Work is needed to extrapolate signals from global climate models, and overlay them with local climate models and data in order to generate reliable projections which can be used to inform local business and policy planning.

1.4 Business Action—Why Is this of Concern for Business, and What Is Business Doing to Address the Challenges?

Water is a necessary input in the manufacture, delivery and use of virtually all products and services. While most companies are water users, other companies are water providers. Companies may use water directly because they need it for their operations (within the product, in the process or simply for consumption/use by employees). They may also depend on water indirectly as a result of water reliance within their value chain (upstream supplier dependencies, downstream consumer dependencies). Companies often also depend on certain habitats (ecosystems) such as forests and uplands that help to regulate the flow and quality of water.

In the early 2000s, it became increasingly clear that a water-constrained world possesses significant risk for businesses, and that all companies are affected by uncertainties, tensions and dilemmas associated with their water consumption. The business community therefore started engaging much more in water management issues and changing their practices and processes in terms of water conservation, wastewater treatment, recycling and reducing carbon footprints. Accompanying business on this journey, the World Business Council for Sustainable Development (WBCSD) started collecting basic information about the current world water situation and identified specific actions that business can take, culminating in two publications, namely *Water Facts and Trends* and *Actions for Sustainable Water Management.* The WBCSD and its member companies then embarked on a scenario project to clarify and enhance business understanding of the key issues and drivers of change related to water, promote mutual understanding between the business community and non-business stakeholders on water management issues, and to support effective business action as part of the solution to sustainable water management.

Close to 200 individuals drawn from business, government, intergovernmental and nongovernmental organizations, explored ways in which the future may or may not be like the past and why traditional forecasting methods are not good enough. In particular, this multi-stakeholder exchange served as a catalyst for joint exploration with other actors in society of how businesses can contribute to sustainable water management. Discussions were based on an increased understanding of the importance of water, and how essential it is to everything in life—food, energy, transportation, nature, leisure, identity, culture, social norms, and virtually all the products used on a daily basis. With population growth and economic development driving accelerating demand for everything, the full value of water is becoming increasingly apparent to all. This consultation resulted in the creation of the Business in the World of Water, *WBCSD Water Scenarios to 2025*, which outline the role of business in relation to the growing issue of water in the world, and which were consequently used to inform the respective management processes of the participating companies. The cross-industry setting helped in learning from other industries/contexts faced with the same or similar issues and challenges. Three parallel stories were developed (Fig. 4):

1.4.1 Hydro—The Efficiency Challenge

With economic development, water demand increases more quickly than population. The resulting stress on water resources is exacerbated by low water-use efficiency, especially in the agricultural sector, caused by factors such as outmoded water systems, poor regulatory enforcement, ineffective price signals, and the lack of incentives for changes in behaviour, particularly by those who claim historical rights to water access. The efficiency challenge in the world of water calls for more value per drop—and "more drops for less," including the value that comes from more jobs per drop, less energy and pollution per drop, efficient water use, and more water for less environmental impact. This efficiency challenge poses an innovation challenge to business—not only in producing new products and services, but also in avoiding or addressing legacy constraints.

Three Parallel Stories

Scenario	Hydro	Rivers	Ocean
Water Challenge	Efficiency (more drops for less and more value per drop)	Security (quantity and quality for all)	Interconnectivity (taking the whole system into account)
Business Challenge	Innovation	Social license to operate	Business role in water governance

Fig. 4 Three parallel stories, WABCD Water Scenarios to 2050

1.4.2 Rivers—The Security Challenge

The increasing stress on local water supplies in many parts of the world raises the issue of water security—quantity and quality for all. The major challenge in water security is to ensure that water is allocated and managed effectively and that there is enough to meet all needs—including the ecosystem services and products on which many livelihoods and economies depend. The political processes for re-allocating water fairly and effectively are of fundamental importance: if local water security is not maintained, the business challenge will include preserving the social licence to operate in that area—even for businesses that are not directly involved in water issues.

1.4.3 Ocean—The Interconnectivity Challenge

Human security and development cannot be isolated from the health and viability of the earth's underlying life support systems. The interconnectivity challenge requires us to be able to think and act in terms of multiple geographies of connection, from nation states and city limits to watersheds and river basins, and in terms of multiple timeframes, in order to ensure that short-term interests do not foreclose longer-term possibilities. The interconnectivity challenge also requires us to take into account not only 'blue water' issues of the water we see, such as the water in lakes and rivers, but also the so-called 'green water' contained in healthy soils, and the 'virtual' or embedded water contained in traded products and services.

1.5 Identifying Business Water Risk—Global Water Tool (GWT)

As a concrete outcome of the scenario work mapping out the water risks and challenges for business and society, the WBCSD's water group identified the need for a practical and easy-to-use tool that could help business understand more clearly their interaction with water resources and serve as a basis for developing corporate water strategies. Twenty-two WBCSD members formed an advisory group, led by CH2M, to develop and provide oversight and pilot testing of the Global Water Tool in order to address these concerns.

The question was not only how much water companies are using, but where. To manage water globally, one needs to know the water situation locally. Water scarcity may mean that even the most efficient operations may be too heavy a burden on local conditions; in the same way, if water is locally plentiful, expensive conservation methods may not be cost-effective. So companies must understand their water needs in relation to the local situation to make intelligent decisions.

By comparing a company's sites with the best available water and sanitation data on a country and watershed basis, the Global Water Tool allows to answer the following questions: How many of my sites are in extremely water-scarce areas? How many of my employees live in countries that lack access to improved water and sanitation? How many of my suppliers will be in water-stressed regions in 2050?

The tool is also linked to Google Earth which provides spatial viewing of a company's site location in relation to detailed geographic information, including surface water and population density.

Concretely, the Global Water Tool:

- Compares a company's **water use** (including staff presence, industrial use and supply chain) with validated water and sanitation availability information, both on a country and watershed basis. However, a volumetric measurement of a company's water use does not accurately reflect water-related impacts or risks. A more realistic idea is obtained by comparing water consumption data with local externalities, including factors that will add pressure on the resource in the future (e.g. population and industrialization growth rates).
- It calculates **water consumption and efficiency**. Comparing water consumption data over a period of time can help a company assess improvements in water use efficiency and monitor progress against sustainable policies or goals.
- It establishes **relative water risks** in a company's portfolio in order to prioritize action. While an overall water reduction goal is simple, the ability to identify sites with the highest water risks can allow for optimal use of funds to generate the biggest impact. The tool allows companies to focus upon the sites requiring the implementation of water management plans and on the suppliers that need support in order to achieve this. It creates key water **Global Reporting Initiative (GRI)** indicators, inventories, risk and performance metrics, as one

cannot manage what one does not measure. The Global Water Tool generates output which can be used for internal water management strategies and goal-setting, as well as for external reporting. It enables **effective communication** between internal and external stakeholders on water issues—companies are increasingly expected to report their sustainability performance and commitments. The Global Water Tool generates key water data and easy-to-understand visual information for effective communication to all stakeholders.

Some companies have used the results of the Global Water Tool to prioritize action for water-stressed plants and support a structured dialogue with sites at risk leading to detailed site level assessments using other tools such as the Global Environmental Management Initiative's *Collecting the Drops* Survey. For example, PepsiCo has included data for corporate-owned manufacturing, distribution, and office sites; international franchise bottler sites; US bottling partner sites. These maps are used to illustrate and communicate water scarcity challenges. Combined Watershed Metrics feature provides snapshots of current and projected stress as a starting point for detailed site analysis.

Others used the results as an input to their internal environmental data-tracking tool to assist with long-term water management strategy. The Global Water Tool can also be used as a strategic tool during the evaluation of new projects. The tool's outputs (e.g. charts and maps) have been used by some companies to improve their communication on water risk, through the development of internal policy and guidelines, but also for reporting purposes, including the recently launched Carbon Disclosure Project's water questionnaire.

Still others used the information obtained from the tool to educate supply chain partners and assess training and capacity building needs. Others still plan to fine-tune publicly stated water goals so that they better align with true regional concerns and risk based on the insights gained from applying the Global Water Tool.

Since its launch in 2007, the Global Water Tool has been updated periodically when new data sets became available and has been used by about 400 global companies. Its widespread applicability has enabled companies from diverse sectors —i.e. from mining to consumer goods—to find value in its utilization. It has been increasingly recognized as the best available approach for companies to assess corporate water-related risks, improve decision-making, shape water management plans, and strengthen communications with internal and external stakeholders on water issues.

Furthermore, specific customizations of the Global Water Tool have been developed for the Oil and Gas, Cement and Power sectors. In addition a customized version, the India Water Tool (IWT) been launched to help water users in India respond to the India specific water challenges by measuring and mapping water risks at a meaningful scale. Since the launch of the India WaterTool in the beginning of 2015, more than 8000 unique visitors to the tool have been recorded in the first six months.

In order to go to the next level of water risk mapping, it is important that companies adopt a value chain approach; often the biggest material risks, impacts and water dependencies are associated with a company's extended value chain, rather than in its direct operations. In addition, companies should focus on your company's material water risks—i.e. important water use impacts (which are not necessarily where the bulk of the water is used)—rather than undertaking a comprehensive assessment of the entire supply chain, or low water volume uses.

1.6 Way Forward—Managing Water as a Shared Resource

Water is a shared resource where the actions of others profoundly impact a company's ability to sustain its own operations. Shared risk is the idea that companies have an incentive to invest in sustainable water management beyond their fence-line in a way that advances the public interest, because it manages business risk at the same time. Shared risk does not imply that water challenges create an equal and similar risk or sense of urgency to all stakeholders. Rather, the concept elevates local water challenges as common or shared problem (shared resource = shared responsibility = shared opportunity = shared risks) and, ideally, triggers proactive and collaborative responses (WWF 2013). Most importantly, shared risk also means shared opportunity.

So it is important to accelerate the business movement worldwide from risk assessment to implementation of response strategies at watershed level in collaboration with other key stakeholders. In essence, more advanced businesses would 'pull' other businesses and SMEs (who often face the greatest water risk) towards best practice to ensure transferability across sectors and across supply chains. Strengthening widespread application of water risk assessment tools and stakeholder mapping to design response strategies at watershed level, will be an essential enabler.

Sound water management allows companies to identify and manage water-related business risks and help communities and governments advance good water governance, sustainable water balance, good water quality and the protection of ecosystems, reducing overall shared risks. However, business cannot do this alone. There is a need for a global drive that water of adequate quality and quantity is shared among all users and is in a more sustainable balance with the limits of renewable water resources and ecosystems requirements. Improved watershed collaboration between key stakeholders to reduce shared water risk is a sine qua non condition to reach this vision. More specifically, local water management must focus on concrete improvements in fields such as:

a. Significant reduction of the amount of untreated wastewater discharge and increase in water reuse;
b. Significant reduction of the rate of agricultural water withdrawals in water stressed aquifers and river basins;
c. Significant acceleration of the progressive realization of the human right to water and sanitation.

1.7 Water Reuse

Reuse answers the challenges of deteriorating quality of raw water sources and increasing variability of availability, as a complement to demand management (imposed restrictions) and desalination (which costs would remain higher than reuse). Large quantities of freshwater can be saved by water reuse and recycling, lowering costs, reducing environmental pollution and improving carbon footprint. Reuse enhances water supply security, which may be endangered in the future due to increased freshwater demands from sectors like agriculture, energy, industry and potable purposes.

Water reuse is becoming a more sustainable and cost-effective alternative water supply, using natural processes as well as more advanced membrane technologies with more efficient energy use. A wide variety of treatment options exist such that health-related risks are considered manageable. Reuse has gone through a rapid curve of innovation and technologies are available. For example, the market volume for membranes has boomed in the last few years.

Water reuse cannot take place in isolation. Communities and businesses need to start managing water differently, bringing competing demands and users together (and adopting an integrated approach to the entire urban area). This includes treating wastewater to appropriate standards for different types of reuses (e.g. cooling, building, mining, irrigation, landscaping, aquifer recharge and potable reuse). There is still a huge potential for recycling and reuse, especially for the reuse of secondary municipal effluents.

Water reuse is of increasing importance not only in arid regions but also in growing cities and contaminated environments. Most of the significant developments in water reuse have occurred in arid regions of the world including Australia, China, Mediterranean countries, Middle East and the United States. But even in temperate regions water reuse is characterized by fast development, in particular for industrial purposes, environmental enhancement and urban recycling.

The limiting factor for water reuse can in many circumstances be the quality of the water available linked to the treatment processes (technology) and potential hazards for secondary users.

Many examples exist where the techno-economic cost benefit would seemingly favour a progression forward on reuse projects. But beyond life cycle assessments that might support a validation of expected return on investment or payback period, commercial considerations (related to service reliability and risk assurance) can be a key factor in a decision to proceed with water reuse.

1.8 Agriculture Water Withdrawals

Increasing demand for food and other agricultural products will put great strains on land, water, energy and other resources in the coming years, and also heavily impact greenhouse gas emissions and climate change. Food demand is expected to rise by 60 % by 2050 due to population growth and increased consumption. Fibre demand for wood panels and paper is predicted to increase by 80–95 % (FAO 2012a). A threefold increase is expected in the demand for biofuels at a time of increasing pressure on water quality and quantity.

Add these projected increases to a world where agriculture is already the world's largest water user, consuming 70 % of total water each year, and energy consumption in agriculture is anticipated to increase by 84 % in 2050 (FAO 2012b) in a business-as-usual scenario. Additionally, climate change will impact food production in several ways. The disturbing result could be increased greenhouse gas emissions, compounded water requirements, decreased yields, and a potential increase in pests and weeds.

Fortunately, an array of "smart" solutions exist and are being developed to usher us towards agricultural production that is knowledge-intensive, more precise and less wasteful, and which utilizes such innovations as smart seeds, clever crop agronomy, zero-energy farms and integrated logistical systems.

These solutions are already available and can be implemented with multiple benefits on yields, energy, water, climate change, land and resource use. Many of these solutions can be "co-optimized" to reinforce each other and deliver multiple synchronized benefits of energy and water savings while increasing yields and creating better quality products. Imagine enzymes that help crops grow faster but also aid in the uptake of fertilizer, saving energy and pollution. Or biodegradable plastic mulch that prevents water loss through evaporation increases soil temperature and accelerates natural nitrogen fixation.

Changing behaviour at the retailer and consumer levels to control food waste would also significantly reduce demand for water and energy embedded in products that never reach an end-user. Co-optimization solutions can go a long way towards addressing climate change mitigation and adaptation. Several solutions do not just improve the productivity of scarce water and energy but also reduce greenhouse gas emissions. Prime examples are the new methods of growing rice that keep methane emissions much lower. Reduced energy consumption in agriculture also immediately affects CO_2 emissions. Meanwhile, better use of water can improve water storage and improve adaptation of climate change fluctuation, and green soil water management adds to improved climate resilience as well.

1.9 Ensuring Safe Water, Sanitation and Hygiene for Employees

Worldwide more than 1.8 billion people depend on unsafe drinking water (Onda et al. 2012) while 4.1 billion lack access to adequate sanitation (Baum et al. 2013); this violates the human right to water and sanitation, and is a major impediment to the economic and social development of millions of households. Furthermore, 1 billion people live in poor urban settlements with limited access to centralized water and sanitation services (UN Water 2010), and by 2050 an estimated 3 billion people will be living in the urban areas of the developing world. Public infrastructure services are not keeping up with existing population growth, and service models to provide sustainable integrated solutions for water and sanitation challenges have thus far failed to reach sufficient scale (Hystra 2011).

Many businesses have operations, employees, contractors and customers in countries lacking access to safe water, sanitation and hygiene (WASH). Economic, social and environmental impacts can cause illness or fatalities, impair productivity, and restrict markets for some products and services. The business cost arising from preventable water-related diseases alone can be material (SIWI 2005). Public health and resilience of local and global economies are seriously affected by these negative impacts on human resources not to mention by the serious environmental damage caused to water polluted by untreated human waste.

Investing in safe WASH for employees provides for healthier and more productive workforce: Adequate access to safe WASH is associated with decreased absenteeism due to water-related diseases, and thus improved productivity. It can also lead to tangible benefits, such as improved public perception and lower reputational risks and more secured social licence to operate.

2 Conclusion

To achieve a more active involvement of the business community in sustainable water management, we need:

- Water governance systems that allow for fair, equitable and sustainable allocation of water for all the stakeholders are a key element of sustainable water management;
- Corporate recognition of the importance of integrated approaches is increasing: shift from cooperating within sectors to collaborating across sectors;
- Corporate shift from only managing water risk to seeing water-related opportunities;
- Corporate shift from taking water for granted to recognizing the true value of water, or more broadly the value of the services provided by the ecosystems.

References

Baum R, Luh J, Bartram J (2013) Sanitation: a global estimate of sewerage connections without treatment and the resulting impact on MDG progress. Environ Sci Tech

Ceres (2009) Water scarcity and climate change: growing risks for business and investors

FAO (Food and Agriculture Organization of the United Nations) (2012a) State of the world's forests 2012. Food and Agriculture Organization of the United Nations. Rome, Italy

FAO (Food and Agriculture Organization of the United Nations) (2012b) World agriculture towards 2030/2050: the 2012 revision. ESA working paper no. 12-03. Food and Agriculture Organization of the United Nations. Rome, Italy
Gleeson T, Wada Y, Bierkens MFP, van Beek LPH et al (2012) Water balance of global aquifers revealed by groundwater footprint. http://www.nature.com/nature/journal/v488/n7410/full/nature11295.html.
Hystra (2011) Access to water for the base of the pyramid. Lessons learned from 15 case studies
IPCC synthesis report (2007) http://www.ipcc.ch/pdf/assessment-report/ar4/syr/ar4_syr.pdf.
Molden D, Frenken K, Barker R, DeFraiture C, Mati B, Svendsen M, Sadoff C, Finlayson M, Atapattu S, Giordano M, Inocencio A, Lannerstad M, Manning N, Molle F, Smedema B, Vallee D (2007) Trends in water and agricultural development. In: Molden D, (ed). Water for food, water for life: a comprehensive assessment of water management in agriculture, Earthscan, London, UK and International Water Management Institute, Colombo, Sri Lanka, pp 57–89
OECD (2012) OECD Environmental outlook to 2050: the consequences of inaction. OECD Publishing, Paris
Onda K, LoBuglio J, Bartram J (2012) Global access to safe water: accounting for water quality and the resulting impact on MDG progress. Int J Environ Res Public Health
Stockholm international water institute (SIWI) (2005) Making water a part of economic development: the economic benefits of improved water management and services http://www.unwater.org/downloads/CSD_Making_water_part_of_economic_development_2005.pdf
UNEP, UN-Habitat (2010) Sick water? the central role of wastewater management in sustainable development. available at: http://www.grida.no/publications/rr/sickwater/.
UN Water (2010) www.un.org/waterforlifedecade/swm_cities_zaragoza_2010/pdf/facts_and_figures_long_final_eng.pdf
Water Resources Group (2009) Charting our water future: economic frameworks to inform decision-making
WHO Global hearth observatory http://www.who.int/gho/urban_health/situation_trends/urban_population_growth_text/en/
WWF (2009) Investigating shared risk in water. http://www.wwf.org.uk/wwf_articles.cfm?unewsid=2836.
WWF (2013) Water stewardship brief, wwf.panda.org

Water, Climate Change and Sustainable Development: An Industry Example

Sanjay Khajuria

Abstract Over more than a decade, Nestlé has engaged in numerous initiatives to create and support climate action actions and accelerate individual and collective efforts to address the potential impacts of climate change. Some of these initiatives include the Business Leadership platform, the United Nations Environment Programme and The Global Compact. Within its sustainability framework, the company has on-going programmes for water and biodiversity conservation, increasing resource use efficiency, no-deforestation, reduction of air pollution emissions, adaptation to climate change and zero waste. Additionally, within its Rural Development Framework, the company has developed its Natural Resource Stewardship with inputs from important stakeholders such as Solidaridad, Rainforest Alliance, The Fair Labor Association and Danish Institute for Human Rights. Many of these initiatives are carried out by Nestlé in India and this chapter includes examples of the initiatives.

1 Introduction

Climate change is increasingly seen as a potent threat to both society and business. The current understanding of climate science suggests that multiple changes can be expected in the future including longer and more frequent heat waves, increased extreme rainfall events and flooding, ocean acidification, and rising sea levels. This will impact economies and societies significantly. Governments and businesses across the world are beginning to understand and appreciate that unless concerted and responsible actions are undertaken urgently we are at serious risk and this will have grave implications. The challenge that looms ahead is immense. Clearly as the IPCC Report indicates "Climate change is a threat to security, food, and mankind".

It is estimated that world population is projected to cross the 8 billion mark in 2028 (United Nations Department of Economic and Social Affairs Population Division 2015) and the middle class is forecast to increase to 5 billion—nearly two

S. Khajuria (✉)
Corporate Affairs, Nestlé India Limited, Gurgaon, India
e-mail: sanjay.khajuria@in.nestle.com

A.K. Biswas and C. Tortajada (eds.), *Water Security, Climate Change and Sustainable Development*, Water Resources Development and Management, DOI 10.1007/978-981-287-976-9_8

thirds of global population (Gertz and Kharas 2010), this coupled with changing lifestyles and eating habits will result in an increasing demand for food. At a time when water is already becoming increasingly scarce, natural resources are constrained and biodiversity is declining, the effect of climate change on weather patterns are further impacting agriculture and influencing crop production and water availability. It has been accepted by the Governments that in order to minimize the risks of the worst impacts of climate change, it needs to stabilize global average warming below 2 °C, and this is possible only if the carbon from manmade CO_2 emissions is restricted to one trillion tonnes. However, it is believed that already over half of this has been emitted and it will cross the trillion tonne mark in less than 30 years (see http://www.trillionthtonne.org/). It is time to step up commitment for rapid and extensive climate action.

The Business Leadership platform has been working with UNEP and The Global Compact amongst other bodies to create support and endorsement for climate action initiatives and accelerate efforts, both individually and collectively, to counter the threat of catastrophic climate change. Nestlé has long maintained that for a company to be successful over time and create value for shareholders it must also create value for society, and consistently focus on sustainable use of resources. Over the last decade or more, Nestlé has participated in industry efforts to help cocoa and coffee farmers adapt to climate change, while also tackling deforestation and helping farmers in dairy farms. Its efforts to further reduce greenhouse gas emissions at factory sites have involved energy efficient programmes, the use of cleaner fuels such as natural gas and biomass and through the Nestlé Environmental Management System it is committed to the continual improvement of the environmental performance of its activities. It is committed to full compliance with all applicable legal environmental requirements and its internal requirements where specific environmental legislation is non-existent or insufficient. It undertakes the systematic assessment and optimization of environmental impacts in the design of new and renovated products; the responsible sourcing of ingredients, packaging materials and other goods and services from suppliers who comply with the Nestlé Supplier Code and demonstrate continual improvement of its environmental performance.

Nestlé is committed to provide climate change leadership and has on-going programmes in several areas that focus on water preservation, natural resources efficiency, biodiversity conservation and no-deforestation, air emissions reduction, climate change adaptation, and zero waste. Natural resource stewardship is a very specific area of work in the Rural Development Framework (RDF) that Nestlé has developed with inputs from specialist stakeholders including Solidaridad, Rainforest Alliance, The Fair Labor Association, and Danish Institute for Human Rights.

2 External Assessments

As a result of the rigorous activities and focus, Nestlé is one of the top food products companies in the 2015 Dow Jones Sustainability Index (see http://www.sustainability-indices.com/). Nestlé scored 99 % in the 'Environment Dimension' of the Index, reflecting the company's strong policies and transparent reporting in the area of environmental sustainability, including its climate change mitigation activities, water management practices and raw material sourcing.

In October 2014, Nestlé received a Climate Disclosure Leadership Index Award from environmental data reporting platform CDP, having achieved a score of 96 out of 100 and it also continues to be among the leading companies for 2015 in Oxfam's sustainability scorecard.

While Nestlé globally undertakes activities which assist in reduction and preservation of water; reducing impact of climate change; improving food security and farmers' incomes by working with farmers and suppliers, this chapter provides examples of the work done by Nestlé in India in these areas directly and with its partners.

Nestlé India has presence across the country with 8 manufacturing locations and 5 co-manufacturing locations. It has 4 branch offices and has its Head office at Gurgaon. Nestlé has been present in India for over 100 years when it started trading in year 1912. It started its first manufacturing facility at Moga in 1961. It is a recognized leader in Nutrition, Health and Wellness and works closely with communities.

3 Water Conservation and Preservation

The World Health Organization (WHO) estimates that every person needs 50–100 L of fresh water per day to ensure that our most basic daily needs are met and few health concerns arise but in 2013, around 768 million people remained without access to clean water (Progress on sanitation and drinking-water—2013 update). With estimated population increase to 2.3 billion by 2050, the situation is likely to get worse. Water scarcity may also lead to 30 % shortfall in global cereal production by year 2030—in less than one and a half decade (see http://www.weforum.org/news/forum-highlights-water-key-factor-food-and-energy-security).

The global water issues have to be addressed by multi stakeholders. Currently 70 % of water is used by agriculture, 20 % by industry and 10 % by municipalities for consumer usage (FAO). For Nestlé, water management is a priority and it works with various stakeholders, including agriculturists, suppliers, international and national bodies, to minimize the impact.

In June 2013, Nestlé renewed the scope of its vision for Water, with the introduction of new Nestlé Commitment on Water Stewardship and in this commitment document, it acknowledged its responsibilities and has outlined the actions needed to be implemented. Nestlé is also the founding signatory of CEO water mandate and the World Business Council for Sustainable Development (WBCSD) pledge for 'Water, Sanitation and Hygiene Implementation at the Workplace'. The 2030 Water Resource Group is chaired by Peter Brabeck-Letmathe, Chairman, Nestlé.

4 Water Saving Initiatives

Nestlé efficiently uses water in its factories around the world, and working towards providing a sustainable supply of water for them is an important priority (Nestlé in Society 2012). It recognizes that this critical natural resource must be used efficiently at its factories and distribution centres, measuring the amount of water used and monitoring its quality upon return to the local watershed. It has set robust water efficiency targets, which continue to drive improvement programmes across all of its operational sites.

Within Nestlé factories and in areas under its control there is a continuous effort to maximize production while minimizing the consumption of natural resources and reducing waste and emissions. It has developed a series of environmental performance indicators (EPIs) in 1997 to monitor its efforts for the sustainable use of natural resources in manufacturing. All processes follow the Nestlé Environmental Management System, and its business practices comply with government policies, environmental laws and regulation.

Some of Nestlé key initiatives to ensure sustainable, environmentally friendly operations are: continuous upgrading of energy and water management practices, reduced water usage in manufacturing, ensuring the absence of pollutants by focusing on zero waste discharge, treating and recycling waste water within the factory and using treated waste water to develop a 'green' environment within its factories.

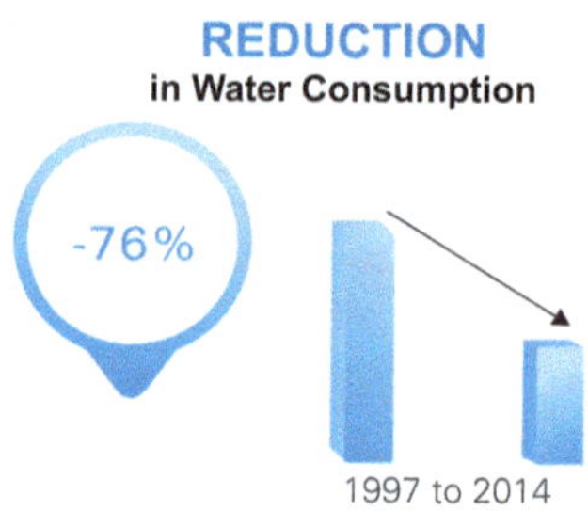

During 1997–2014, Nestlé in India reduced water consumption per tonne of production substantially by around 76 % through key initiatives taken in the last many years such as: reduction in cooling water consumption with the help of latest technology in Chiller cooling towers, reuse of waste water after treatment in Industrial Services Cooling Towers, optimizing the water consumption in process lines and adopting new cleaning—in—process (CIP) technology (Nestlé India Limited 2014).

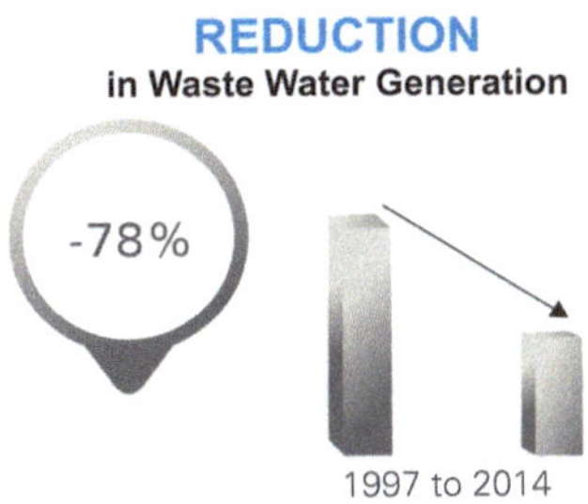

During 1997–2014, Nestlé has reduced wastewater discharge per tonne of production by around 78 % through many initiatives: using more efficient cleaning processes to reduce its water utilization, reusing water after Reverse Osmosis treatment and recovering steam condensation for reuse in process (Nestlé India Limited 2014).

5 Saving Water Through the Zer'Eau Initiative

Nestlé has recently implemented Zer'Eau (zero water), an initiative to help improve water availability in the community in water-scarce areas. Responding to the water scarcity and the need for enhanced water conservation, it has accelerated water use reduction projects and initiatives including at its Moga factory in India. The Project involves recovery of condensate from milk for processing; thus reducing dependency on external sources of water. The recovered water can be used for the boiler, utilities and other applications.

6 Water Access and Sanitation in the Community

As mentioned, the WHO estimates that every person needs 50–100 L of fresh water per day to ensure that our basic daily needs are met and few health concerns arise. Nestlé ensures that its operations do not compromise the right to water of local communities, provide access to clean water and sanitation for employees, and assist in the provision of clean water and improved sanitation to priority communities adjacent to its factories where it sources its agricultural commodities (see http://www.nestle.com/csv/water). Nestlé globally provides access to water, sanitation and hygiene projects to 350,000 beneficiaries around our manufacturing locations and in farmer connect areas.

In 1999, Nestlé in India carried out a pilot survey in village communities in Moga, where one of its main factories is based. The study highlighted a lack of clean drinking water as the key concern for the community. Many village schools were only equipped with hand pumps which provided poor quality of drinking water to their students.

To help improve access to safe drinking water, Nestlé initiated construction of clean drinking water facilities in schools around all of its factories to benefit the surrounding communities. Prior to constructing a drinking water facility in a particular school, it assesses the number of students, population of the surrounding community, their proximity to a water source, and the school/community's financial capabilities. This assessment is done in order to ensure that the project benefits a sizeable number of people who do not have access to alternate drinking water sources and who cannot afford to construct such facilities themselves. The sourced water is stored in hygienic tanks enclosed in a specially designed facility to preserve the quality of the water. It involves the school and surrounding community through joint ownership of the water tanks, which helps to ensure better upkeep and maintenance of the tanks. From the very first project setup in 1999, village communities have been encouraged to contribute voluntarily to these projects. Villagers take pride in the project and their ownership of it, making this 'joint ownership' a success.

Students are encouraged to form 'Water Committees' for the maintenance and upkeep of the tanks. However, water samples are tested regularly for potability in Nestlé laboratories. Till 2014, Nestlé India has built 223 water projects across the country, benefitting over 94,000 students in village communities.

The benefits are visible. *In the words of Mr. Gurdarshan Singh, Principal, Government High School, Village Khosa Pando, "With the bore well facility provided by Nestlé school students have access to drinking water at all times during the day."*

Nestlé conducts Water Awareness Programs for students at the schools where it builds drinking water facilities. Through this programme, it aims to make water ambassadors and create awareness amongst students regarding water conservation and protection of water resources to ensure the responsible utilization of water for a sustainable future.

Students are taught through posters and demonstrations on water saving and purification methods such as the Drip method, Solar Water Disinfection Process (SODIS), Rain water harvesting model, etc. Water committees comprising of school children are set up in these schools to propagate judicious water consumption and ensure proper upkeep and maintenance of the project. Water saving tips on the water storage tanks reinforce key messages on water conservation. Over 51,000 students had successfully participated in Water Awareness Programmes till end of December, 2014.

As Ravinder Kaur, village Talwandi Bhangeria stated, "*Now our children come home and explain to us how to make proper use of water. For example, while brushing teeth one should use a cup of water rather than letting water taps flow freely. This will lead to better water management in our village community.*"

As a result of surveys carried out in rural areas, lack of sanitation facilities was found to be one of the main reasons for discouraging girls from attending school. In an effort to promote the UN Millennium Development Goals of Universal Primary Education and to ensure availability of sanitation facilities, Nestlé constructs sanitation facilities for girl students in village schools to help reduce the dropout rate among girl students. So far the company has built about 100 sanitation facilities across India benefitting over 36,500 girl students. This initiative has reduced dropouts in schools.

7 Access to Water and Sanitation in Our Operations

Nestlé is one of the first signatories of the pledge drawn up by the World Business Council for Sustainable Development (WBCSD) that commits businesses to upholding the human right to water and sanitation within their operations (see http://www.wbcsd.org/washatworkplace.aspx). The pledge for 'Water, Sanitation and Hygiene Implementation at the Workplace' (WASH) challenges companies to provide access to safe water and sanitation, and appropriate facilities to ensure personal hygiene, to employees in all premises in their direct control, within three years of signing. By 2016, Nestlé's objective is that globally 350,000 beneficiaries will have access to water, sanitation or hygiene projects around our manufacturing facilities and in farmer connect areas.

8 Employee Engagement

Through Nestlé Continuous Excellence initiatives, Nestlé develops the methods and tools for its employees. The initiative fosters a systematic, employee involved, continuous improvement culture to develop environmentally sustainable business practices. Employees are trained and educated which enables them to make well informed decisions on water stewardship.

9 Reduction in Energy Consumption

Nestlé factories transform perishable raw materials into safe and nutritious food products and are continuously exploring opportunities to improve efficiency and environmental performance. Some of the initiatives Nestlé focuses on pertain to optimizing energy; Minimize waste generation; use sustainably managed renewable energy sources; Recover value from by-products; and Control and eliminate emissions. Energy conservation remains a key focus area as it is a significant part of the environmental management of Nestlé factories. Each year, energy reduction objectives are set and the factories implement programmes and innovative initiatives to save energy. These include engineering interventions, processing initiatives, raising employee awareness, training, measuring and monitoring usage and energy auditing.

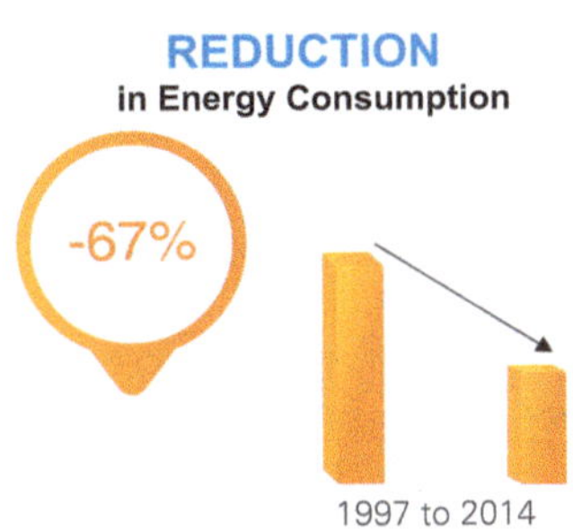

During 1997–2014, Nestlé in India reduced energy consumption per tonne of production substantially and improved energy use efficiency by around 67 % through multiple initiatives (Nestlé India Limited 2014). Some of the key initiatives being: tea waste generated from processing of green tea leaves is used as fuel for steam generation along with effluent sludge, continuously review energy to track and replace energy inefficient equipment, investment in processes to reduce energy losses like, installation of Variable Frequency Drivers on high capacity motors, auto operations and controls, process modification to reduce energy utilization such as, installation of Wipe Film Evaporator.

Also innovatively using the waste heat of one process as input for another e.g. exhaust heat of the generator is used to produce steam for manufacturing, optimize the steam consumption in de-aerators, reduce steam losses in steam lines, optimize the electrical load of utilities and recover flash steam.

Nestlé has set up solar panels to provide hot water for cleaning and washing to over 900 milk collection agencies in the milk shed area of Moga, Punjab.

Nestlé encourages farmers to set up biogas plants as an alternate source of energy. Biogas plants are among one of the main projects promoting decentralized power solutions, by the Ministry of Non-Conventional Energy Sources (MNES) (see http://mnre.gov.in/schemes/decentralized-systems/schems-2/). Biogas is a clean, non-polluting, low-cost fuel, which can be produced from cattle dung, human waste and other organic matter.

10 Greenhouse Gas Emissions

Intergovernmental Panel on Climate Change (IPCC) shows that global emissions of greenhouse gases have risen to unprecedented levels despite a growing number of policies to reduce climate change. Emissions grew more quickly between 2000 and 2010 than in previous decades (IPCC WG III). A range of policies, including those on climate change, energy security, and sustainable development, have been effective in reducing GHG emissions in different sectors and many countries. These measures, however, have not been significant enough to impact the global growth in emissions. Increasing levels of anthropogenic greenhouse gases (GHG) in the atmosphere are causing climate changes. Fossil fuel usage and industrial processes are estimated to contribute 78 % of Green House Gases and need to be looked at on priority (IPCC IV 2007). Two other areas are deforestation and dairy.

Nestlé believes that it is imperative that governments, companies and individuals take measures to reduce the production of greenhouse gases. They each have different responsibilities and roles: Governments provide stable legal and policy frameworks that create a level playing field and provide incentives to guide behaviour to tackle climate change; companies compete to produce the most energy efficient products and services and increase climate resilience with their suppliers; and consumers choose the most sustainable products. Nestlé itself aims to constantly lower the GHG emissions associated with the production and distribution of its food and beverages, and to design products that help consumers lower their own GHG emissions in the use of those products.

Nestlé helps to protect the environment and scarce resources, now and for future generations by improving efficiency in its own operations and involving its partners to continuously optimize the environmental performance of its products along the value chain.

Nestlé in India is also cutting emissions by shifting from road transportation to rail transportation which helps cut traffic congestion on roads and leads to lesser air pollution.

Nestlé has encouraged 400 of its biggest suppliers to report their GHG emissions though Climate Disclosure Project (CDP).

11 Zero Waste

The impact of food waste is not just financial. Environmentally, food waste leads to wasteful use of chemicals such as fertilizers and pesticides; more fuel used for transportation; and more rotting food, creating more methane—one of the most harmful greenhouse gases that contributes to climate change. Methane is 23 times more potent than CO_2 as a greenhouse gas. The vast amount of food going to landfills makes a significant contribution to global warming.

Roughly one third of the food produced in the world for human consumption every year—approximately 1.3 billion tonnes—gets lost or wasted.

Every year, consumers in rich countries waste almost as much food (222 million tonnes) as the entire net food production of sub-Saharan Africa. The amount of food lost or wasted every year is equivalent to more than half of the world's annual cereals crop (2.3 billion tonnes in 2009/2010). Food loss and waste also amount to major squandering of multiple resources including land, water, energy, labour, financial resources and contributes to climate change (UNEP).

As per estimates, India loses close to USD 8 billion worth of perishable agricultural communities for various reasons including lack of processing. The solution could lie in strengthening the supply chain by supporting farmers and investments in infrastructure, transportation, as well as in an expansion of the food processing industry. Efficient infrastructure and packaging could also contribute to reducing the amount of food waste.

Nestlé continually finds ways to reduce food waste throughout the value chain. Avoiding food waste at various stages of the value chain gives a great opportunity to save natural resources. It has been helping to avoid food wastage by transforming perishable raw materials such as milk, fruits, vegetables etc. into safe, nutritious and tastier value-added food products.

In India alone, Nestlé procures over 1.5 million tonnes of perishable raw materials into food products for consumers every year. It also assists famers to improve productivity. In the milk supply chain, where it engages with close to 100,000 farmers in India, it has provided cooling facilities to farmers that have cut milk loss between farm and retail—saving water, energy and GHG emissions as well as money.

Nestlé takes steps to continually improve its demand planning. This reduces food waste by supporting manufacturing efficiency; reducing wastage due to age, obsolescence or the inconvenient location of finished goods and increasing freshness at the point of sale.

12 Support to Farmers/Suppliers

With the impact of rising global temperatures already visible, Nestlé also believes that a new emphasis needs to be made by Governments, businesses and other stakeholders to increase efforts in helping farmers become resilient to weather changes. There is a general need for the provision of better information and insights for all stakeholders on climate change adaptation, plants that are better adapted for changing climate, and help for farmers to adopt better agricultural practices and landscape management techniques.

The Nestlé Supplier Code ("the Code") defines the non-negotiable minimum standards that it asks suppliers and their sub-tier suppliers to respect and to adhere to when conducting business with Nestlé. This document helps the continued implementation of Nestlé commitment to international standards such as the OECD Guidelines for Multinational Enterprises, the UN Guiding Principles on Business and Human Rights, the Core Conventions of the International Labour Organization (ILO) and the 10 Principles of the United Nations Global Compact, beyond its own operations, to every link of its upstream supply chain back to farms and plantations. The Code is an extension of our Nestlé Corporate Business Principles and the foundation of our Responsible Sourcing programme.

The standards of the Code set forth expectations for suppliers with whom Nestlé does business, as well as all others with whom they do business including all employees, upstream suppliers and other third-parties. It includes adherence to all applicable laws and regulations and strive to comply with international and industry standards and best practices. In addition, through its Responsible Sourcing program, Nestlé verifies compliance with the Code through internal and external assessment mechanisms. It continuously encourages suppliers to improve their operations. In case of improvement required, or direct sourcing from smallholders or farmers, it supports in the establishment of milestones and systems to ensure that practices are continuously upgraded.

Environmental Sustainability is among the 4 pillars of the Supplier Code. Through the Code, Nestlé requires its suppliers to comply with all applicable legal environmental requirements and demonstrate continual improvement of their environmental performance. It fully supports environmental permits and reporting, environmental management system, hazardous materials and product safety and resource consumption, pollution prevention and waste minimization.

The Responsible Sourcing Guideline (RSG) (Nestlé) complements the Nestlé Supplier Code and is applicable to all stages of the upstream value chain back to the primary production level. The Guidelines apply to all materials of agricultural, forestry, fishery and aquaculture origin. The Nestlé Commitment on Deforestation and Forest Stewardship details its commitment to both tackle deforestation and improve the standard of forest stewardship, through the responsible purchasing of products from forests and forested landscape.

13 Coffee Farmers—NESCAFÉ Plan

The NESCAFÉ Plan is one such example of Nestlé's support. It is a global initiative that brings together under one umbrella, Nestlé's commitments to support responsible farming, production and consumption. To Nestlé, goodness is more than just providing the best quality in a cup- it goes 'Beyond the cup'.

In India, Nestlé kick started the NESCAFÉ plan implementation in 2012, with an aim to share our global coffee expertise with coffee farmers in the coffee growing regions of South India and train them in responsible farming to improve their agricultural practices to improve quality, productivity and sustainability over the long term. For this, it has a set up a coffee demonstration farm and training centre in Coorg. It also supports farmers in conducting assessments and obtaining certification from the Common Code for the Coffee Community (see http://www.4c-coffeeassociation.org/) (4C Association). In 2012, Nestlé established three 4C units at Kushalnagar, Kalpetta and Mudigere. In line with the its' ambition to make coffee farming an attractive and sustainable activity for future generations, over 1200 Coffee Farmers have been trained in NESCAFÉ Better Farming Practices and provided with technical assistance. The training sessions help the farmers to increase efficiencies in coffee productivity and quality, optimize costs and improve social and environmental impact. The coffee farmers are provided with technical assistance on eco-friendly waste collection and disposal practices in the coffee farms, adopting soil and water conservation practices and eco-friendly chemical waste water treatment units, adopting measures to use water sustainably, plant protection measures like wearing mask, gloves and plastic over coats during chemical spray and reducing the use of non-renewable energy. These initiatives collectively help coffee farmers to better adapt to climate change and environmental challenges.

In 2014, Nestlé undertook a wide scientific research on soil status and components to adapt accordingly at farm level, through soil testing facilities. Farmers were given fertilizer recommendations to help then save money and increase their margins.

It also organized Occupational Health and Safety Programmes for coffee farmers to assess health and safety conditions to 4C Coffee Farmers, workers and their family members were provided with free of cost health services including, first aid care and specialized check-ups on diabetes, dermatology, children health, ECG, dental care, ophthalmology, etc.

Shri Jawaid Akhtar, I.A.S, Chairman of the Coffee Board of India stated at the Launch of the programme "I am happy that the NESCAFÉ PLAN is being launched in India. It will provide coffee farmers with technology and best practices for sustainable production of high quality coffee and also benefit them with improved access to markets."

14 Dairy Farmers

Nestlé relationships are based on the understanding that its investments must result in long term sustainable growth and economic and social progress for the community as well. Thus in addition to collecting milk and implementing its milk district model successfully to ensure a stable livelihood for local dairy farmers, Nestlé also supports the sustainable development of its farmers by assisting them to increase milk productivity and expand their herds via financial assistance, technical assistance regarding feeding practices, breeding, mechanization of dairy farms and making veterinary services available through our team of around 35 veterinarians, promotion of sustainable agricultural practices, farmer training and education, promote silage making, balance feeding and ensuring best quality concentrate feed mix and improving animal breed with imported semen.

Today, Nestlé works with around 100,000 milk farmers and collects over 300 million kg of high quality milk every year across the States of Punjab, Haryana and Rajasthan.

15 Helping Consumers to Reduce GHG Emissions with Information and Tips

Nestlé acknowledges the interest from consumers to know about the impact of food products during their entire life cycle from farm to consumer and beyond. Nestlé is committed to meaningful and accurate environmental information and dialogue based on scientific evidence about our products, activities and services at corporate, market and product brand level. Nestlé is therefore helping consumers make informed choices through readable, substantiated communication; leveraging relevant contact points (e.g. digital, packaging and point of sale) to inform consumers of environmental improvements, as well as action they can take when using our products and handling used packaging.

The labelling of Nestlé Products has been enhanced with a QR code that allows consumers to use their mobile phones to find out more about the company and the product they hold in their hands. On scanning the QR code on the label, consumers can get information specific to the product in India under three heads, which are Nutrition, Environment and Society. These also provide simple tips to the consumers to help them enjoy food and beverage products while saving money and taking care of the environment.

16 Further Thoughts

Nestle uses its scientific expertise and resources to help find ways to improve crop resilience through its R&D centres. It help produce disease and drought resistant varieties. Nestlé continues to partner with experts in the field of water to develop awareness and helps implement best practices regarding water conservation and sustainability.

The study "Measuring the water footprints of milk production: contributions to livelihood benefits and sustainable water use in the Moga District in Punjab" was conducted in partnership with the International Water Management Institute (see http://www.iwmi.cgiar.org/publications/other-publication-types/reports/). It researches the main factors for groundwater depletion in the Moga district by studying the water footprint for agriculture, and recommends interventions for sustainable irrigation and agriculture. These recommendations are being promoted by Nestlé among the dairy farmers in the region.

In summary, the issues of climate change and water scarcity have taken centre stage. Changes to the climate are already occurring. There is a threat to food systems and food security. India is water stressed and changes to the climate will impact production of main commodities. On an individual basis and at multi stakeholder level, adaptation efforts are becoming a priority, however, more can be done. Nestlé is making efforts across value chain from farmer to consumer, in line with its focus on water, rural development and nutrition. It is helping farmers and suppliers adapt to changing weather patterns. It is looking at energy saving programs at its factories; water conservation at factories, working with school children and making them water ambassadors, farmers and suppliers; and in reducing GHG gases. It remains committed to continual improvement of environmental performance of its products and services and the activities in factories and across its value chain. It also recognizes that farmers income, specially small farmers needs to be protected and makes efforts in training and providing support to them. It invests in measures to increase the resilience of its operations and work with suppliers and small holder farmers be it in dairy or coffee. Nestlé is also making consumers a partner in these initiatives by providing social and environmental information so that they can make informed choices. Communications tools such as NESCAFÉ Life Cycle Assessment and Nestlé Beyond the Label will continue to be a focus to provide product information to consumers. Nestlé's efforts to reduce food wastage along the value chain also contribute to reduce the emissions.

References

FAO (United Nations Food and Agriculture Organization) Growing more food—using less water, http://www.fao.org/fileadmin/user_upload/newsroom/docs/water_facts.pdf

Gertz G, Kharas H (2010) The new global middle class: a cross-over from West to East. Wolfensohn Center for Development at Brookings, http://www.brookings.edu/research/papers/2010/03/china-middle-class-kharas

IPCC WG III: Greenhouse gas emissions accelerate despite reduction efforts, https://www.ipcc.ch/pdf/ar5/pr_wg3/20140413_pr_pc_wg3_en.pdf

IPCC IV (2007) IPCC Fourth assessment report: climate change, https://www.ipcc.ch/publications_and_data/ar4/syr/en/contents.html

Nestlé in Society (2012) Creating shared value and meeting our commitments, http://www.nestle.com.au/asset-library/Documents/Nestle-CSV-Summary-Report-2012-EN.pdf

Nestlé India Limited (2014) Business responsibility report, https://www.nestle.in/investors/stockandfinancials/business-responsibility-report

UNEP (United Nations Environment Programme) Environment for development: food waste facts, http://www.unep.org/wed/2013/quickfacts/

United Nations Department of Economic and Social Affairs Population Division (2015) Expert group meeting on policy responses to low fertility. New York, http://www.un.org/en/development/desa/population/events/expert-group/24/index.shtml

Journey of Sustainable Development by Private Sector Actors

Asim Parekh

Abstract While valiant efforts have been made to reduce natural resource extraction and companies are increasingly incorporating sustainability initiatives into business planning, greater collective action, integrated government policies, innovation and support are needed to ensure natural resource sustainability—especially when it comes to water, the heart of sustainable development. After exploring three key water trends of the last 30 years, this chapter looks ahead to the next 30 years, informing of three areas where businesses, in partnership with governments and civil society, should focus efforts to make a positive difference on global water challenges.

1 Introduction

The intellect of the human race has developed rapidly in the past 100 years with a corresponding increase in our ability to create value and turn a profit (Vital 2013). But it's humanity's swift and negative impact on natural resources of which we should be in greater awe.

The world economy today uses around 30 % fewer resources to produce one Euro or Dollar of GDP than 30 years ago; however overall resource use is increasing (Friends of the Earth Europe 2009). Humans today extract and use around 50 % more natural resources than only 30 years ago, at about 60 billion tonnes of raw materials a year. Water is one of these precious resources that humans are using at a greater rate. Globally, water use has grown at more than twice the rate of population increase in the last century. This increased use of water is affecting every continent and an increasing number of regions are chronically short of water (United Nations). At least 663 million people lack access to safe water and another 2.4 billion don't have access to improved sanitation (WASH Advocates 2015) leading to increased disease rates and suppressing development. As water chal-

A. Parekh (✉)
Coca-Cola India Private Limited, Enkay Towers, Udyog Vihar,
Phase V, Gurgaon 122016 India
e-mail: aparekh@coca-cola.com

A.K. Biswas and C. Tortajada (eds.), *Water Security, Climate Change and Sustainable Development*, Water Resources Development and Management, DOI 10.1007/978-981-287-976-9_9

lenges escalate, greater water stress and impacts will be felt by our environment, global health, and economic growth.

The time to embark on and embrace the journey of sustainable development was yesterday, is today, and will be tomorrow. While our natural resource extraction has been intense over the last 30 years, so have select efforts towards reducing that usage. Unfortunately, they haven't been enough. Greater collective action, integrated government policies, innovation and support are required to ensure long-term resource availability and efficiency.

2 The Evolution of Sustainable Development

For many companies, sustainable development has evolved from being reputation management to an essential component of their core business strategy. Sustainability is now considered critical to company's growth, brand relevance and

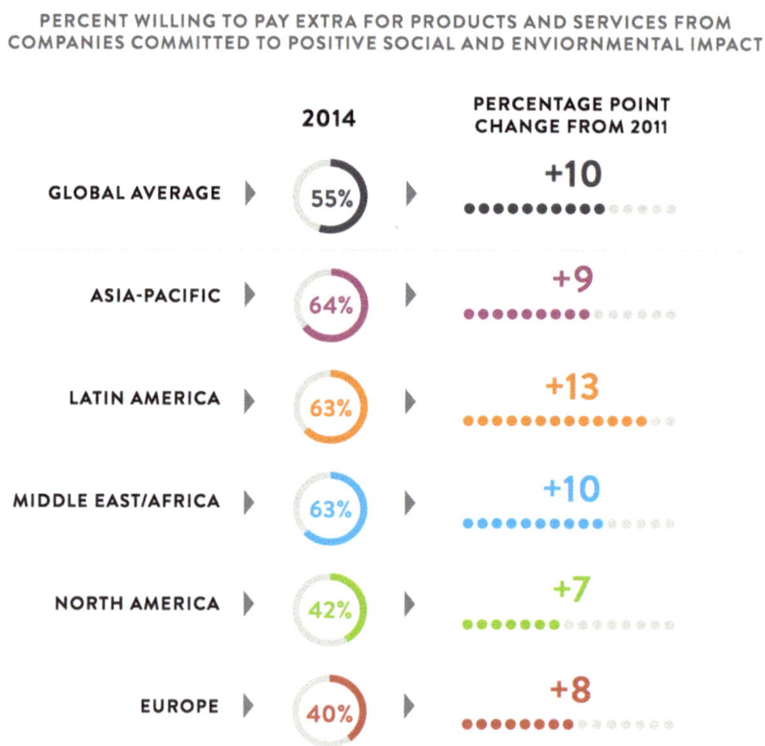

Fig. 1 Of global respondents to Nielsen's corporate social responsibility survey, 55 % are willing to pay extra for products and services from companies committed to social responsibility. *Source* Nielsen's report, Doing Well By Doing Good

long-term value, as well as reputation. In addition to sustainability efforts helping conserve natural resources, today's consumers demand that businesses act in a socially responsible manner.

In a June 2014 report, *Doing Well By Doing Good*, Nielsen confirmed that not only do consumers increasingly care about corporate social responsibility, they're willing to pay for it. The study found that 55 % of respondents would be willing to pay extra for products and services from companies committed to positive social and environmental impact (Fig. 1).

A successful company views sustainability as an enabler of growth and as a market differentiator. In the food and beverage industry, for example, sustainability drives top-line growth through consumer preference for sustainably-sourced ingredients and development innovation, and bottom-line growth through areas such as waste reduction and efficiency improvement, creating economic value and reducing impacts on the planet (Fig. 2). In the longer term, sustainability becomes a must for business as they seek to ensure continuity of inputs and supplies (whether that be land, fresh water, fuel, etc.). In other words, companies will have to work to sustain natural resources in order to survive, let alone grow.

Coca-Cola has determined that when people learn about the Company's commitment to sustainability, they're more likely to be Coca-Cola advocates. Similarly when consumers know about Coca-Cola sustainability efforts, trust increases.

The United Nations Global Compact's *Global Corporate Sustainability Report 2013* found that companies committed to the UN Global Compact (2013), the

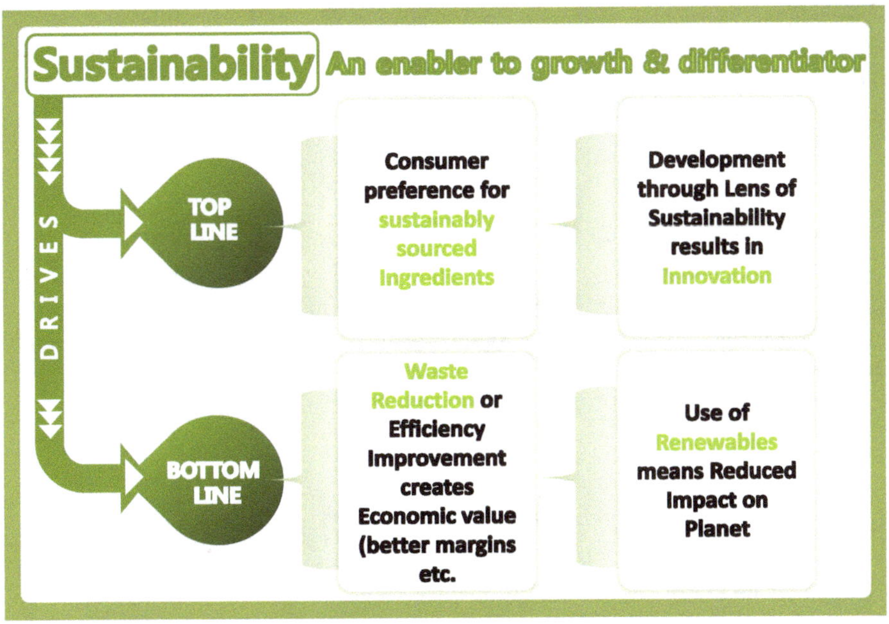

Fig. 2 Sustainability as an enabler to growth and differentiation. *Source* Internal

world's largest corporate sustainability initiative, are moving from good intentions to significant actions. Companies indicated an understanding that addressing sustainability issues is both good for business and the communities where they operate. The report cautions that while progress is being realized, there is a long journey ahead for companies to fully embed responsible practices across their organizations and supply chains.

3 Sustainable Development and Water

When it comes to sustainable development, water is at the heart of its success. However, water scarcity will be a major issue faced by many communities around the world this century. Humans, ecosystems, and economies all require water for survival, yet climate change, population growth and overuse of natural resources are presenting water challenges. According to UN Water, "around 1.2 billion people, or almost one-fifth of the world's population, live in areas of physical scarcity, and 500 million people are approaching this situation. Another 1.6 billion people, or almost one quarter of the world's population, face economic water shortage, where countries lack the necessary infrastructure to take water from rivers and aquifers."

While the planet's fresh water is finite, it is infinitely renewable. The world contains enough water to meet individual, ecological, agricultural and business needs, but only if everyone works to better manage water resources. Improved water management, and relevant water projects that make a difference, require cross-sector collective action—from business, civil society and government. But before we can understand where to direct our action, we must revisit where we've been.

4 Three Key Water Trends of the Last 30 Years

When examining the past 30 years, three important trends surface related to addressing water challenges: (1) Establishment of the United Nations Millennium Development Goals; (2) Media, Investor, and Civil Society Focus on Water Issues; and (3) Corporate Engagement.

4.1 Millennium Development Goals

Established in 2000, the United Nations (UN) Millennium Development Goals (MDGs) for 2015 established global recognition of social and environmental challenges, as well as a commitment to make significant progress. The goals on improving access to safe water and sanitation were critical, not only on their own but in their relationship to the other goals, which depend on these basic human needs being addressed to have any chance of success.

The WHO/UNICEF Joint Monitoring Programme (JMP) cited that the goal to halve the proportion of people without access to improved water sources had been met, though there is much debate as to the data, as well as the sustainability of interventions for the long-term. The JMP also noted that most countries were well behind on meeting sanitation goals. With mixed progress on these goals, their establishment was nonetheless significant as it considerably raised awareness in the developed world, generated funds and action (in the developed and developing worlds), and drove governments to invest and act. The achievements have considerably benefitted the lives and livelihoods of billions of people, including 2.3 billion people who gained access to improved drinking water resources between 1990 and 2012 (JMP, http://www.wssinfo.org/).

4.2 Media, Investor and Civil Society Focus on Water Issues

Though much of the past 30 years has focused on climate change science, effects and mitigation, water has clearly risen to the top of concerns and focus across the world. The impacts of climate (i.e., adaptation) are recognized as being mainly in water. The media has focused reporting on water issues, crises and responses, which has amplified public concern and driven action across sectors.

Investors in public companies followed suit (and in some cases led the media) to an acute awareness of water scarcity and pollution issues and their associated impact to bottom lines, further prompting action, and importantly gaining corporate C-suite attention. Companies now list water and other environmental concerns as risks in reporting. In 2003, The Coca-Cola Company was among the first to report water quality and quantity as a material risk to its business in its U.S. Securities and Exchange Commission Form 10-K for investors, and continues to include it as a risk today. Many food and beverage companies, including Dr Pepper Snapple Group and PepsiCo, also report environmental risks.

Civil society (e.g., NGOs, CBOs) has seen an explosion of water issue-related actions and new organizations, funding drives, awareness campaigns, and advocacy. As an example, 2014 saw World Wildlife Fund (WWF) recognize freshwater conservation as a top priority. WWF partners with communities, businesses and others to decrease pollution, increase water efficiency and protect natural areas to ensure enough clean water exists to conserve wildlife and provide a healthy future for all.

In fact, the organization has placed a major emphasis on transforming business, recognizing that all companies rely on renewable natural resources, yet their operations can have a significant negative impact on the environment. WWF engages in strategic and innovative partnerships with companies to help reduce their environmental footprints; champion pre-competitive and sustainable solutions; and harness the global marketplace as a force for conservation.

WWF partners with Coca-Cola to address the natural resource challenges that impact fresh water, with Avon Products to protect the world's forests, and 30 leading companies on climate protection initiatives through their Climate Savers programme. Other nonprofits, including The Nature Conservancy and Rainforest Alliance, have also realized the benefit of public-private partnerships and collaborate with business.

Adeeb Mahmud (2014), director of FSG, says that NGOs are playing three distinct new roles in partnerships with companies: innovation partner, vendor with a mission, and strategic advisor. NGOs often possess the technical skills, local relationships and knowledge of local challenges that can make or break new market introduction or market staying power (The Guardian). In FSG's report, *Ahead of the Curve* (Peterson et al. 2013), shared-value partnerships are highlighted as an opportunity to create greater impact.

4.3 Corporate Engagement

Though very few corporations publicly recognized and considered water stewardship a critical, business imperative early on, and subsequently reacted with risk mitigation plans, many corporations now have reacted to the growing awareness and realties of issues, as well as pressure from shareholders, government and the public.

The private sector, especially the food and beverage industry, is clearly present for policy discussions, to lend thought leadership, and in support of science and action. Much more needs to be achieved, particularly in shared supply chain with agriculture and livestock, but efforts by World Business Council for Sustainable Development (WBCSD 2012), CEO Water Mandate and others clearly demonstrate the wide adoption across business.

CEO-led WBCSD encourages members to work across sectors, geographies and value chains to help explore, develop and scale up business solutions to address the world's most pressing sustainability challenges. In 2012, WBCSD developed a step-by-step guide, *Collaboration, innovation, transformation*, to help businesses accelerate sustainable growth in their value chains. The how-to guide features case studies from 10 leading companies, including Proctor & Gamble, The Coca-Cola Company and Unilever, demonstrating how sustainable value chains can create competitive advantage for business.

In 2007, the unique public private initiative, CEO Water Mandate, was introduced to assist companies in the development, implementation and disclosure of water sustainability policies and practices. Companies that endorse the mandate are required to report progress annually against a set of standard principles. More than 100 companies, such as Danone, General Mills, PepsiCo, The Coca-Cola Company, have endorsed the mandate.

Through the 2030 Water Resources Group (2030 WRG), key public decision-makers, concerned private sector champions and civil society representatives are working to help improve water policy in several countries around the world. At the invitation of governments, 2030 WRG helps water officials accelerate reforms aimed at ensuring sustainable water resource management for the long-term development and economic growth of their countries. 2030 WRG helps countries first diagnose gaps in their water supplies, and then develop and test solutions. Private sector involvement includes Nestlé, The Coca-Cola Company, PepsiCo and SABMiller.

As part of The Coca-Cola Company's global partnership with WWF, partners set a goal to integrate the value of nature into public and private decision-making processes. This work includes advancing the science to fill gaps and establishing the business case for incorporating biodiversity and ecosystem services into corporate decision-making processes.

5 Corporate Engagement: Coca-Cola Working at the Heart of Sustainable Development

With more than 250 bottling partners across more than 200 countries, Coca-Cola believes it has a particular obligation and a unique opportunity to be sustainability stewards. Unifying its expansive system is a shared sustainability vision framed as "Me (people), We (communities), and World (environment)". Through this framework, water, women and well-being have been prioritized because of the important role each plays in Coca-Cola's business and the Company's ability to make a positive difference in these areas (Fig. 3).

Fig. 3 The Coca-Cola Company's "Me, We, World" global sustainability framework

Under the aligned framework, sustainability commitments were established and are intended to be met by 2020. Some of those commitments are to introduce product variety where Coca-Cola operates, to economically empower 5 million women entrepreneurs, and to replenish 100 % of the water used in finished products.

Water, which is at the heart of sustainable development, is a top priority for Coca-Cola because it is an essential ingredient in all of the Company's beverages and needed to produce the agricultural ingredients on which it relies. Water also is critical to the health and economic prosperity of the communities it serves. If the communities Coca-Cola serves are not sustainable, neither is the business.

Backing this belief, Coca-Cola set a goal to 'give back' an amount of water equivalent to what is used in beverages and their production by 2020. To achieve this goal, the company is focused on improving water-use efficiency, treating all wastewater from manufacturing operations, and replenishing 100 % of the water used in finished beverages back to communities and nature through the support of healthy watersheds and community water programs. In addition, Coca-Cola requires bottling partners to assess the quality and quantity of water sources and to develop and implement an associated comprehensive Source Water Protection Plan (SWPP).

Key global progress highlights include:

- Improved water efficiency 10 % since 2010, and nearly 25 % since 2005 (Fig. 4);
- Aligned 99 % with Company wastewater standards (Fig. 5);
- Replenished an estimated 94 % of the water used in finished beverages, a total of 153.6 billion litres, through 209 community water projects in 61 countries with key partners (Fig. 6);
- Source vulnerability assessments (SVA) were completed and SWPPs have been completed or are in development for all of our bottling plants globally.

With agriculture accounting for approximately 70 % of worldwide water withdrawals, Coca-Cola has also committed to sustainably sourcing key agricultural ingredients for its products.

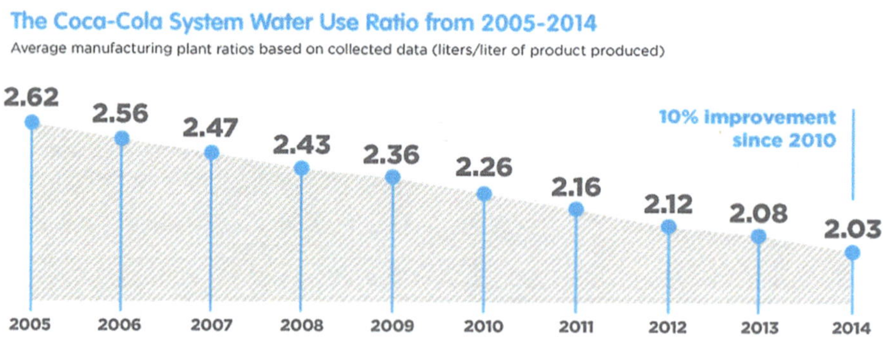

Fig. 4 The Coca-Cola system global water-use ratio 2005–2014

The Coca-Cola Company Requirements for Treated Wastewater Quality*

(liters/liter of product produced)

		Maximum Value (unless applicable legal requirements are more stringent)
5-Day Biological Oxygen Demand	<50 MG/L	
pH Level	6.5-8**	
Total Suspended Solids	<50 MG/L	
Total Dissolved Solids	<2,000 MG/L	
Total Nitrogen	<5 MG/L	
Total Phosphorus	<2 MG/L**	

* These are six of the 20 water quality parameters established for the Coca-Cola system.

** Depends on receiving stream water conditions.

Fig. 5 Coca-Cola's requirements for treated wastewater

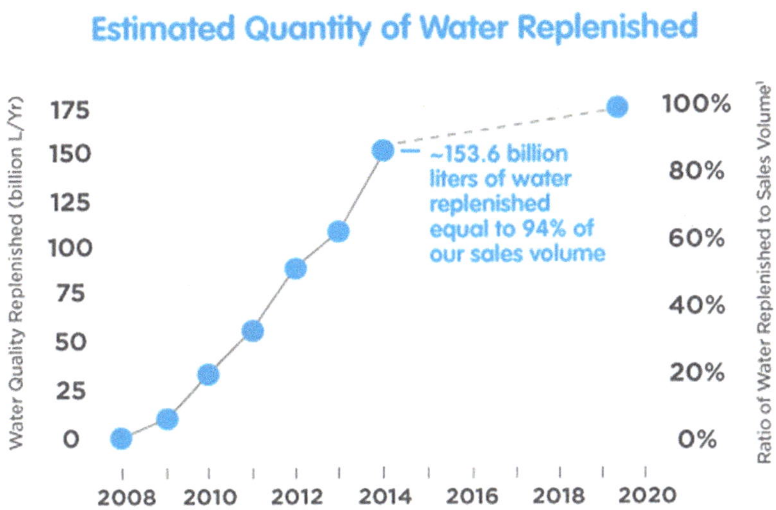

Fig. 6 Coca-Cola's quantity of water replenished from community water partnership (CWP) projects

Ingredients include cane sugar, beet sugar, high-fructose starch-based syrup (primarily corn), tea, coffee, palm oil, soy, pulp and paper fibre, oranges, lemons, grapes, apples and mangoes. Conservation partner WWF is assisting Coca-Cola with the implementation of its Sustainable Agriculture Guiding Principles throughout the Coca-Cola system.

5.1 Coca-Cola India Snapshot

For Coca-Cola, global water stewardship progress requires commitment and dedication from partners around the world. A standout is the Coca-Cola system in India, where full balance has been achieved between groundwater used in beverage production and that replenished to nature and communities. Local bottling plants are continuously working to identify opportunities to reduce consumption, recycle and reuse water. With NGOs, the government, other industries, farmers and local communities, Coca-Cola has been able to identify priority areas and implement projects to improve local water conditions.

Replenish groundwater efforts are focused on rainwater harvesting, constructing check dams, restoring ponds and other natural bodies, and supporting agricultural improvements. Coca-Cola has approximately 382 projects for creating water replenishment potential across 22 of India's 35 states and territories with the estimated potential to return nearly 130 % of the groundwater it uses (Fig. 7).

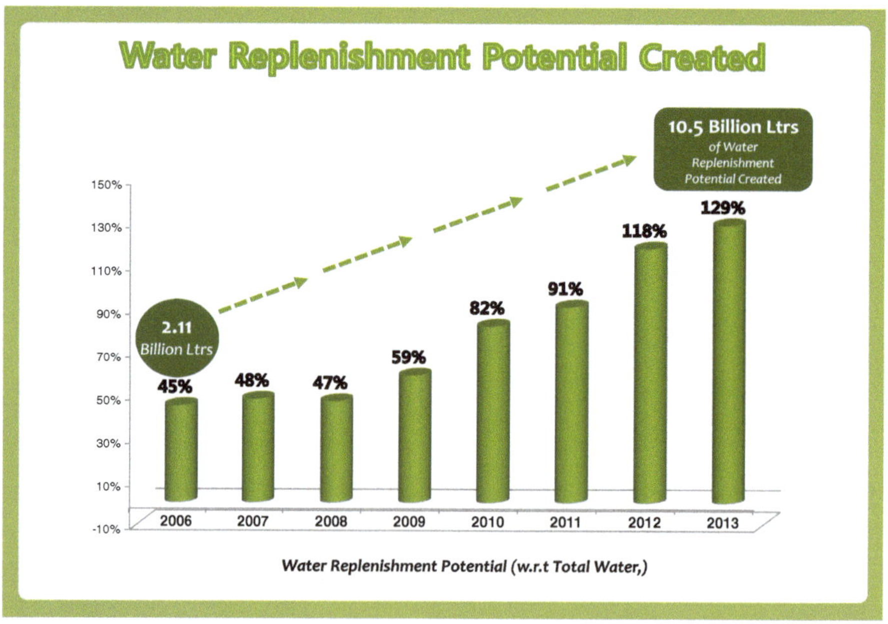

Fig. 7 Coca-Cola India has the potential to replenish nearly 130 % of the groundwater it uses

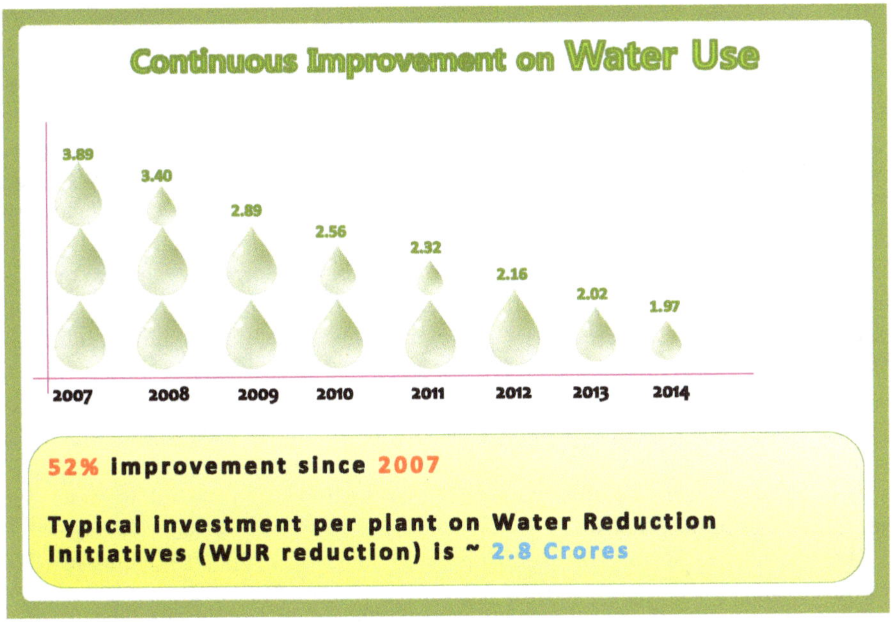

Fig. 8 Coca-Cola India's water-use improvement

The Company has been involved in more than 500 drip irrigation projects in partnership with farmers and government agencies, to improve the efficiency and productivity of water used for irrigation. Also with farmers, laser-levelling projects on 2715 acres of farmland have been undertaken, saving an estimated 3 billion litres of groundwater per year.

From 2007 to 2014, the water-use ratio (water required to produce a litre of beverage) has improved by over 52 %. In 2007, the water-use ratio (WUR) was 3.89 L/L and improved to 1.97 L/L in 2014. This significant improvement enables the production of more beverages using less water and also maintaining the quantity of treated effluent discharge (Fig. 8).

Coca-Cola India's innovation and commitment to sustainable development extends beyond water stewardship to sustainable agriculture, climate protection and renewable resources.

6 Further Food and Beverage Industry Engagement: Nestlé, PepsiCo and SAB Miller

Food and beverage peers and partners have also embraced the sustainable development track with water as a key component of their initiatives. As an industry, with a global and local presence, the collective capability to truly make a positive difference towards a healthier planet is immense.

Nestlé is a peer often recognized for water stewardship efforts (2013a). In 2013, the company launched the *Nestlé Commitment on Water Stewardship*, a document to guide activities towards the sustainable management of shared water resources. As part of that guide, Nestlé is working to achieve water efficiency across its operations, with a 2015 goal to reduce direct water withdrawals per tonne of product in every product category to achieve an overall reduction of 40 % since 2005. By 2013, Nestlé reduced direct water withdrawals in every product category, achieving an overall reduction per tonne of product of 33 % since 2005 (Fig. 9).

PepsiCo's sustainability agenda "Performance with Purpose," has a goal to deliver sustained financial performance and is built on three sustainability pillars: human, environmental and talent. Under environmental sustainability, in 2012, PepsiCo met its target to partner to provide access to safe water to 3 million people in developing countries by the end of 2015, three years ahead of its goal (Fig. 10). Access to safe water was achieved at watershed, community and household levels by making water more readily available, better managing supply or volume of water, and/or ensuring quality through water treatment, improved hygiene and community sanitation (PepsiCo).

Water resources are one of SABMiller's Sustainability Shared Imperatives. For its breweries, the beer and soft drinks business set a target to reduce water use 25 % by 2015. In March 2014, the company met this target early, crediting the efforts of its breweries to drive improvements throughout manufacturing processes. In 2014, SABMiller used 621 million hl of water to produce beer, compared to 667 million hl in 2013 (Fig. 11).

Another performance and forward-thinking highlight is a report SABMiller launched with WWF at the 2014 World Economic Form. The report, *The Water—*

2013 water withdrawals per product category

Product category	**Water withdrawal (m^3/tonne)** 2005	2013
Powdered and liquid beverages	22.91	9.85
Water	1.96	1.58[12]
Milk products and Ice cream	8.33	4.77
Nutrition and HealthCare	19.34	11.02
Prepared dishes and cooking aids	7.12	4.22
Confectionery	9.22	4.7
PetCare	1.44	1.05
Total	4.41	2.92

Fig. 9 Nestlé's water withdrawals per product category. *Source* Nestlé in society: creating shared value and meeting our commitments 2013b

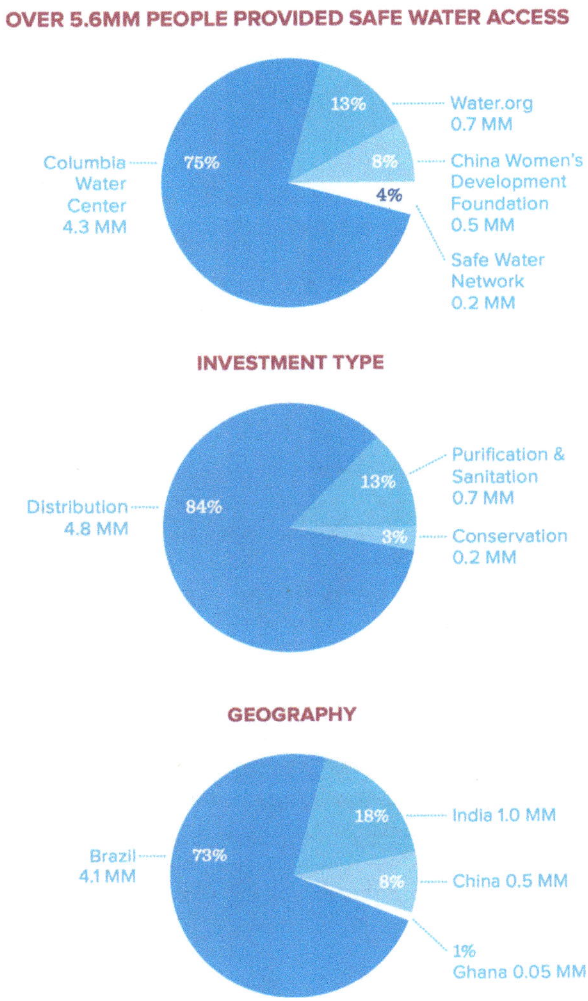

Fig. 10 PepsiCo's access to safe water results. *Source* PepsiCo's water report delivering access to safe water through partnerships (2014)

Food—Energy Nexus: Insights into resilient development, examines 16 countries or states, comparing the ways in which their development patterns have managed their different mixes of resources and different capacities to make use of those resources. With results concluding that the most resilient economic systems combine robust infrastructure, flexible institutions and functioning natural capital, the report's reflective case studies propose how policymakers should respond to nexus challenges.

All of these companies are engaged in many more sustainability initiatives inside and outside of their walls. The industry is coming to recognize that while it must be efficient within its manufacturing operations it must also be effective beyond the property line, helping build sustainable communities. Those communities host and support companies and also grant the social licence to operate, essential to enabling operation over the next 5, 10 and 30 years.

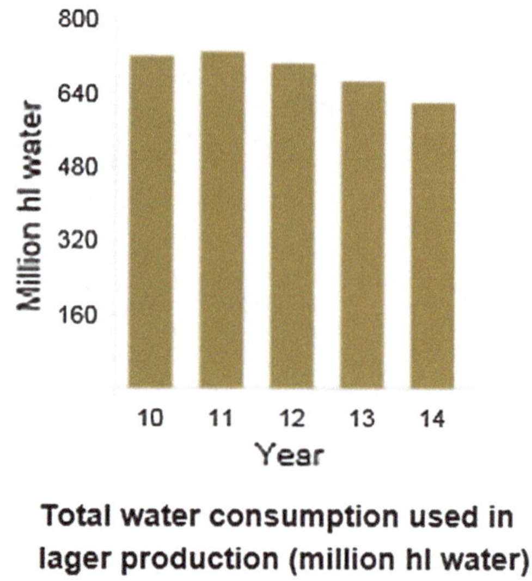

Fig. 11 SABMiller's total water consumption used in lager production. *Source* SABMiller

7 Looking Ahead to the Next 30 Years

By 2030, the world is predicted to have a 40 % deficit between global water supply and demand. It's a significant gap to fill, which is made ever more challenging by the complex nature of water and its relationship with climate change, population growth and rising energy needs. That means less available water for drinking, agriculture, transportation, manufacturing, sanitation and habitats for a diverse range of plants and animals, critical to healthy ecosystems. Working collaboratively, businesses, governments and civil society can make a positive difference on the world's water challenges. Where to focus is the ultimate question. Here are three answers.

7.1 Food-Water-Energy Nexus

The food-water-energy nexus is quickly becoming the defining sustainability challenge for the future. The relation of the three and their interdependences can no longer be ignored in a world of growing populations, rising incomes, urbanization, and climate change.

Gone are the days of abundant water, land, and energy resources. And gone too is single variant thinking to what are now multi-variant problems. Addressing the food-water-energy nexus will require a transformation in education, policy and governance (corporate and government), planning, and cooperation. The food-water-energy nexus will reveal the tradeoffs necessary to meet energy, food and water needs, and tough choices will have to be made by societies, governments and business.

As cross-sector partners focus on water projects, they should aim to help communities mitigate and adapt to the impacts associated with the intersections of water, energy, and food. For example, projects could plan to increase the ability of watersheds to absorb threats associated with increasingly severe weather events and help build resilience in response to the higher demands for water, energy, and food. Partners could target increasing crop yields to meet the needs of a growing population while reducing the impacts on water sources through water body alterations, aquifer recharge, rainwater harvesting and more. Or, efforts could be supported to reduce energy demands by promoting local water sources that eliminate the need to treat and transport water, which is energy intensive.

Water stewardship efforts must extend over company plant walls to local water sources and beyond to agricultural supply chains, helping and encouraging suppliers to improve efficiency and reduce their water use.

7.2 *Climate Change Adaptation*

With already recognized changes in weather patterns, droughts, floods, storm intensity and unpredictability as the new reality in today's changing climate, the world will have to continue and amplify efforts of climate change mitigation.

The absence and impairment of healthy, connected, natural ecosystems (e.g., mangroves and wetlands) can exacerbate the negative impacts of storms and other intense water events. Preserved ecosystems can help weather changes and adapt more readily. The priority climate change adaptation action should be to re-double and expand current efforts related to watershed protection and conservation. Yes, there is a lot of uncertainty but more ecosystem conservation will certainly help and is needed.

One of the United Nations Development Programme's (UNDP 2012) focus areas for climate change mitigation and adaptation is ecosystems and biodiversity, recognizing that nature-based solutions are needed in helping vulnerable communities increase their resilience. Large-scale rehabilitation projects can build natural buffers against disasters amplified by climate change, and can contribute to reducing greenhouse gas emissions (UNDP).

7.3 *Post-2015 Sustainable Development Goals*

Building on the MDGs and converging with the post-2015 development agenda, Sustainable Development Goals (SDGs) are being cultivated. Now is the time to lend a voice to goal creation. When it comes to water security, SDGs need to target universal access to safe drinking water by 2025, at the latest. For the provision of access to improved sanitation, a universal coverage target should be set, with universal access to basic sanitation by 2020 and improved sanitation by 2030.

Freshwater withdrawals must be brought into line with sustainable supply (natural renewal minus environmental flows) by 2030. Adequate treatment (secondary treatment at a minimum) of all municipal and industrial waste water prior to discharge must be achieved by 2030.

Importantly, all development and sustainability goals should take the perspective of the food-water-energy nexus. The world faces challenges and constraints with each of these, highly integrated resources. Isolated goals (e.g., water alone), that do not take into account effects on the other resources (i.e., food and energy) will not work effectively, and well-intentioned efforts to make progress on one goal could hinder progress against others.

8 Get on Board with Sustainable Development

While it may sound simple, it must be said. For our businesses to be sustainable we must act in sustainable ways. There's no sitting on the sidelines while others execute. We must all strive to be more efficient in our operations and more engaged in our communities, viewing community challenges as our own, and meeting water issues with innovative solutions.

We may have come a long way, from sustainability initiatives being viewed as a courtesy to now a critical and strategic part of business planning and execution, but the world still has a long way to go. Strategies and initiatives must be implemented, improved, and increased. Peers must be inspired to join the sustainability journey. Only with everyone on board will we be able to achieve meaningful and positive changes.

References

Adeeb Z. Mahmud (2014) Beyond charity: three innovative types of business partnerships for nonprofits [Blog post]. Retrieved from http://www.theguardian.com/sustainable-business/ngo-business-nonprofit-partnerships-gsk-nespresso-wwf-coke

Friends of the Earth Europe (2009) Overconsumption? Our use of the world's natural resources [Report]. Retrieved from https://www.foe.co.uk/sites/default/files/downloads/overconsumption.pdf

Nestlé (2013a) Nestlé Commitment on Water Stewardship [Appendix]. Retrieved from http://www.nestle.com/asset-library/documents/library/documents/corporate_social_responsibility/nestle-commitment-water-stewardship.pdf

Nestlé (2013b) Creating Shared Value [Report]. Retrieved from http://storage.nestle.com/Interactive_CSV_Full_2013/files/assets/basic-html/page185.html

PepsiCo (2014) Delivering Access to Safe Water through Partnerships [Report]. Retrieved from https://www.pepsico.com/docs/album/water/pep_wp14_safe_water_2014.pdf?sfvrsn=2

Peterson, Kyle, Adeeb Z. Mahmud, Rebecca Yael Weissburg (2013) Ahead of the Curve [Report]. Retrieved from http://futureingo.org/

SABMiller (2014) Building brands, brewing value [Report]. Retrieved from http://www.sabmiller.com/docs/default-source/investor-documents/reports/2014/financial-reports/annual-report-2014.pdf?sfvrsn=10

SABMiller and WWF (2014) The Water-Food-Energy Nexus: Insights into Resilient Development [Report]. Retrieved from http://www.sabmiller.com/docs/default-source/investor-documents/reports/2014/sustainability-reports/water-food-energy-nexus-2014.pdf

The Nielsen Company (2014) Doing Well By Doing Good [Report]. Retrieved from http://www.nielsen.com/content/dam/nielsenglobal/apac/docs/reports/2014/Nielsen-Global-Corporate-Social-Responsibility-Report-June-2014.pdf

UNICEF and World Health Organization (2015) 25 Years Progress on Sanitation and Drinking Water—2015 Update and MDG Assessment [Report]. Retrieved from http://www.wssinfo.org/fileadmin/user_upload/resources/JMP-Update-report-2015_English.pdf

United Nations (2015) We Can End Poverty—Millennium Development Goals 2015 and Beyond [Information from a page]. Retrieved from http://www.un.org/millenniumgoals/environ.shtml

United Nations Department of Economic and Social Affairs (2007). International Decade for Action 'WATER FOR LIFE' 2005-2015, UN Water [Information on a page]. Retrieved from http://www.un.org/waterforlifedecade/scarcity.shtml

United Nations Development Programme (2012) The Future We Want [Report]. Retrieved from http://www.undp.org/content/dam/undp/library/Environment%20and%20Energy/biodiversity/UNDP-Biodiversity-and-Ecosystems-Global-Framework-2012-2020.pdf

United Nations Global Compact (2013) Global Corporate Sustainability Report [Report]. Retrieved from https://www.unglobalcompact.org/library/371

Vital A (2013) The History of Creating Value–How Humans Made Money in Different Eras [Blog post]. Retrieved from http://fundersandfounders.com/the-history-of-creating-value/

WASH Advocates (2015) WASH 101 [Information on a page]. Retrieved from http://www.washadvocates.org/learn/wash-top-facts/

World Business Council for Sustainable Development (2012) Collaboration, innovation, transformation [Guide]. Retrieved from http://www.wbcsd.org/Pages/EDocument/EDocumentDetails.aspx?ID=14257

Will Rivers Become a Cause of Conflict, Rather Than Co-operation, in South Asia?

Adam Roberts

Abstract South Asia faces among the most severe shortages of fresh water anywhere in the world. That scarcity will worsen amid growing demand because of economic and demographic growth and uncertainty over supply, partly as a result of climate change. In a region with a history of cross-border conflict and feeble regional institutions, some analysts warn competition over water supplies, notably rivers that cross borders, is most likely to lead to conflict between countries. This chapter argues resource scarcity need not automatically lead to conflict–abundance can equally be a cause of conflict. How political and military actors choose to respond to scarcity, seeking co-operation or an excuse for conflict, will do more to determine utcomes.

1 Introduction

The term "rivals" is so routinely used to describe relations in South Asia it is easy to forget its meaning. India and Pakistan, separated at birth, are rivals in military, security, diplomatic and other realms. India is an emerging rival to China, in demographic, economic and military terms. Rivalry between India and Bangladesh comes into public discussion over the control of the shared border. Nepal, wedged between India and China, is an arena in which strategic rivals compete.

South Asia is notably prone to competitive, even violent, rivalry. Its regional institutions are undeveloped and ineffective by global standards. The South Asian Association for Regional Co-operation is hardly active, only modestly influential. Trade levels within South Asia are more miserly than within any other region. Levels of investment between neighbours are similarly low. For those who frequently cross the land border between India and Pakistan, or fly between countries of this region, it is obvious how officials make it difficult for their own people—tourists, traders, investors, those attending conferences, journalists and even diplomats. Hostile visa regimes, poor infrastructure, expensive and infrequent flight

A. Roberts (✉)
The Economist, New Delhi, India
e-mail: adamroberts@economist.com

A.K. Biswas and C. Tortajada (eds.), *Water Security, Climate Change and Sustainable Development*, Water Resources Development and Management, DOI 10.1007/978-981-287-976-9_10

connections are all off-putting. The most obvious indicator of rivalry is the region's history of violence: previous wars, terrorist strikes, religious strife, insurgencies and other forms of cross-border clashes are potentially a guide to future trouble.

In that context, this chapter considers another potential element for rivalry, meaning violent conflict or its threat. This relates to the term rival itself, from the Latin "rivalis", originally meaning "person using the same stream as another", from "rivus", or stream. How serious is the threat that competition for access to fresh water, in particular rivers that cross sovereign borders, could escalate into violent disputes?

The focus here is on the possibility of conflict between countries. In passing, however, it is worth noting that water conflict within countries could prove to be the greater problem. Fierce competition for supplies already occurs within sovereign borders. India is replete with examples. In 2014 that country saw the formation of a new state, Telangana, born from the immense Andhra Pradesh. Immediately there arose bitter negotiation and popular protests on how the new upper-riparian state, Telangana, would allow water, especially in the Krishna River, to flow to farmers in the residual state of Andhra Pradesh. Karnataka and Andhra Pradesh have also long disputed how much water can be taken from the Krishna. The state governments of Kerala and Tamil Nadu oppose each other over the safe use of an irrigation dam, over a century old, at Mullaperiyar on the Periyar River. Rulers of the national capital territory, Delhi, have accused an upstream, neighbouring state, Haryana, of diverting excessive water from the Yamuna River, exacerbating water shortages and environmental problems in India's capital. Similar problems exist elsewhere. Inside Pakistan, lower-riparian provinces, such as Sindh, accuse large, politically strong and heavily farmed Punjab province, of extracting more than its fair share of water from the Indus River.

Internal competition for water could, in theory, lead to violent conflict within a particular country. Such a conflict could acquire international dimensions. In general however there exist legal and political systems within South Asian countries to resolve such disputes peacefully. By contrast, internationally, despite existing water treaties and mechanisms for specific areas, there are limited legal and other mechanisms for preventing or resolving conflict. For that reason, the focus here is on competition for fresh water as a source of international rivalry.

2 The "Water Bomb"

Many worry that competition for fresh water is a likely source of war in the region. The head of one think-tank in Mumbai, the Strategic Foresight Group, Sundeep Waslekar, has suggested there exists a "mega arc of hydro insecurity" stretching from China in the east, westwards along northern India, Pakistan and the Himalayas, through Afghanistan to the Middle East. A study by his group, assessing 20 broad issues that might impinge on global security in the coming half century, concluded that competition for scarce water was especially threatening. It foresaw a high likelihood of problems arising as countries compete for water, and speculated that a particularly high impact would result from clashes over water. His greatest concern was the prospect of India and upstream China clashing over fresh water: the single-most "challenging relationship" in this region (South Asia's water 2011).

Others emphasize potential for water conflict between India and Pakistan, compounding an already problematic relationship. The late Boobli George Verghese, an Indian expert on water, suggested in 2011, while based at the Centre for Policy Research in Delhi, that "water is the new battle cry for the jihadis" in Pakistan. In off-record conversations serving and retired senior Indian diplomats and other officials say Pakistan's government raises competition for fresh water as a means to another end: to keep alive an issue of contention to justify future conflict, in case the Kashmir dispute were ever settled. By contrast in Islamabad ministers, diplomats and others say, apparently sincerely, that bilateral rivalry will intensify because of Indian upstream extraction of water from various tributaries to the Indus River. As in much of South Asia, conspiracy theories abound. A senior Kashmiri geologist, in Indian-run Srinagar but reflecting typical Pakistani anxieties, once suggested India's broad strategic intention in Kashmir is to build enough dams to retain sufficient water, so "they will switch the Indus off, to make Pakistan solely dependent on India. It is going to be a water bomb."

In Bangladesh officials, activists, politicians, diplomats and others say relations with India are most troubled when they touch on the border and water sharing. The head of the Bangladesh Institute of Peace and Security Studies, in Dhaka, Major General Muniruzzaman, suggested in one interview that "water is the most critical issue now. If ever there were a localized conflict in South Asia it will be over water. Bangladesh as the lowest riparian state has taken the brunt of it so far. We live with two water hegemons, China and India." Water experts in Bangladesh assess the potential for large-scale dam building in Indian Arunachal Pradesh and ask how the flow of rivers will moderate downstream for good or ill (potentially this is beneficial to mitigate floods). An Islamist politician in Dhaka once suggested that "India plots against its small neighbour on water; this should be raised in international forums". A moderate, liberal, newspaper editor in the same city, accused India of mistreating Bangladesh by building a barrage on the Ganges River at Farakka, which he said unfairly diverted water, with serious economic and environmental consequences downstream which he likened to an "atomic disaster". Tariq Karim, one of Bangladesh's foremost experts on water and on the 1996 Ganges Water Sharing Treaty, also its high commissioner in Delhi, once observed how Pakistan raises the idea of conflict over water as a "tactical weapon" against India. He believes the region would be best served by a collaborative approach to water sharing.

As elsewhere in the world, those in lower riparian countries typically worry most about the behaviour of those upstream. Less frequently an upstream power objects to activity downstream, for example as China discourages India from constructing hydropower projects in Arunachal Pradesh, where the two countries contest sovereignty.

3 From Scarcity to Something Worse

Divergent voices warn of conflict, all beginning with the fact that fresh water is increasingly scarce. By one estimate in 2011 Indians, per person, had access to just 1730 m^3 of fresh water a year, less than a quarter of the global average of 8209. An average Pakistani had much less, with 1000 m^3. China faces a similarly dire situation: it is home to a fifth of the world's population, but only 7 % of the fresh water.

Averages mislead. In India much rain falls in the north-east and south, but the largest and growing population is in the relatively parched centre and north. Yearly averages also hide the fact that water comes unevenly during a year, much in the monsoon. Scarcity in particular regions, at particular times, can be extreme.

Rising demand for water will mean more scarcity. South Asia's population is growing. India is already 1.25 billion-strong, expands by about 1.2 % yearly, or 15 m people, and is unlikely to peak before about 1.7 billion mid century. India will easily become the most populous country in the world. Pakistan is the sixth-largest, with some 190 m, rising at 1.7 % yearly. The United Nations predicts the region will be home to 2.1 billion people by 2040.

Average incomes are also rising, which is associated with rising demand for water per person. This is likely to be seen first in agricultural patterns. By far the greatest share of water consumption (90 % in India, 94 % in Pakistan, 88 % in Bangladesh) goes to farming, with less to industry and households. Farmers, responding to rising demand for more water-intense products, including for export, will seek more water, for example in growing wheat and rice, or in feed for animals. At the same time industrial, construction and residential demand for water will soar.

Rising demand for energy is another factor. A huge expansion of hydropower production is under way in South Asia. In Nepal, by one estimate, enough rivers flow to generate 40 GW of electricity: so dozens of new projects are being built, notably by Indian investors. Four big new ones, alone, could earn Nepal $17 billion over the next three decades, from exports of electricity. In Indian-run Kashmir large hydropower projects, for example at Baglihar and at Kishanganga, are being built. India plans dozens in Arunachal Pradesh. Similarly in Pakistan only 10 % of hydro potential has yet been tapped, according to the late John Briscoe, an expert who long worked with the World Bank. That figure will surely rise. China is also racing to build hydropower projects on the Tsangpo (Brahmaputra) River, that flows into India and eventually Bangladesh.

Dams built for irrigation permanently extract water from rivers so have the most effect downstream. For example Brahma Chellaney has claimed that China's government plans an immense (though scarcely plausible) engineering project to divert the Tsangpo to send water eastwards to thirsty, heavily-populated, eastern China. Most new hydropower projects, however, are "run of the river" systems, which only temporarily divert rivers to spin turbines, releasing water back downstream. Even so such dams do cause small, temporary diversions, or require the filling of small reservoirs. Given large numbers of hydropower projects in one area, or droughts, then the collective effect downstream could potentially be large, adding to scarcity.

Compounding the problem are worries about overall supply. The depletion and pollution of groundwater stocks is common in much of South Asia. In both India and Bangladesh, for example, widespread use of electric or diesel agricultural pumps for irrigation has lowered the water table, at times by hundreds of metres. The same extraction is the likeliest explanation for the heavy presence of arsenic in water pumped in both countries, which has exposed millions to potential poisoning. Immense deposits in rivers of untreated sewage, industrial effluents, agricultural waste and other pollutants, for example in the River Ganges, reduce the supply of water fit even for agricultural ends.

Climate change is an additional concern. Scientists say glaciers are already melting in the Himalayas, known as the Earth's "third pole", which recharge rivers in much of the subcontinent. To blame are rising temperatures but also more soot and other particles blown on to the ice. Faster melting in the coming decades will first mean more water deposited into rivers, but subsequently smaller glaciers will mean reduced flow of some rivers, especially outside of the monsoon.

A changing climate could also make the annual monsoon less predictable. Three-quarters of India's rainfall comes in the monsoon months of June to September. Studies already suggest it is erratic four years in ten. Temperatures in India have been rising over the past 60 years, the risk of instability is rising. One study has found, in the three decades to 2009, a 4.5 % decline in monsoon rainfall (India's climate 2012). R. Krishnan, of the Indian Institute of Tropical Meteorology, has suggested there is evidence of a "steady decline" in monsoon rain in one part of south India. Patterns remain uncertain, but potential damage is great to South Asian agriculture—still largely rain-fed, not irrigated—if rainfall becomes more erratic or diminished.

4 Getting Beyond Scarcity

Those who talk of Asian "water wars" or of freshwater as "Asia's new battleground" are right to say scarcity is already a problem and is likely to worsen. A harder job is explaining how that, in turn, causes conflict. Within India a forceful proponent of the idea of water wars is Brahma Chellaney, who in his 2011 book, "Water, Asia's New Battleground" (Chellaney 2011), worries that competition over water "could become a trigger for war or diplomatic strong-arming". He predicts for example that "Water scarcity is set to become Asia's defining crisis by midcentury".

This is a bold prediction, however a host of other worries in Asian affairs could turn out to be even more awful (nuclear proliferation, the spread of religious militancy and so on). And even if one accepts that competition for finite resources will become extreme, it is imaginable that finite resources aside from water could be a more direct cause of conflict. Diminishing deposits of oil and gas, or scarce quantities of arable land, could provoke clashes. A world in which air is dangerously polluted, and clean air scarce, could see two neighbours clash over dangerously high emissions that cross borders. Chellaney avers that, as a scarce resource, water is "the most critical one, for which there is no substitute". Strictly, even that is not true. Given abundant supplies of cheap energy, desalination of sea water is possible, as oil-rich Gulf States show today by growing (at extraordinary cost) wheat beside the desert. In the Maldives, in 2014, the risks of desalination became evident, when the local plant burned down and the country required India to ship bottled water urgently. But other solutions (if at high economic cost) also exist, such as the piping of fresh water long distances. In practice the competition for fresh water is indeed likely to grow more intense, but it is not self-evident that its scarcity will be the worst crisis of all.

Scarcity of something, after all, can be a cause of co-operation, whereas a time of abundance can be a moment for violence. A study of nomadic groups in recent decades in northern Kenya suggested periods of high rainfall there coincide with violent raids and warfare; whereas times of water shortage are associated with less violence, perhaps because the benefits of raiding and war are reduced (see http://www.tandfonline.com/doi/abs/10.1080/13698240903403915#.VOHMZy61erY). Elsewhere an abundance of easily extractable and valuable resources has frequently been associated with violence. Academics and others have long discussed the idea of a "resource curse" in Africa, for example, holding that immense deposits of mineral resources and more, help to explain why conflict has been so prevalent, state institutions weak and economies distorted.

It is possible, in other words, to imagine that as governments and other actors in South Asia grow more conscious of the problem of scarcity of water, then co-operation, as much as conflict, could follow. After all, India and Pakistan were able to forge the Indus Waters Treaty just over half a century ago, and they continue to abide by its mechanisms and arbitration processes today. India and Nepal have signed three agreements over rivers, in 1954, 1959 and 1996. India and Bangladesh in the 1990s agreed the Ganges Water Sharing Treaty and, though progress is slow, are co-operating on water-sharing on other rivers, such as the Teesta. Around the world there have been some 400 treaties concluded on the use of rivers. Globally there is increasing co-operation on tackling climate change, creating international mechanisms to limit the spread of pollution and to establish norms of behaviour in dealing with shared environmental resources. It is possible to imagine South Asians becoming more deeply involved in these international structures and, in turn, co-operating more closely on related issues, such as the use of fresh water. Similarly India and Nepal are rapidly improving bilateral relations by co-operating on the use of Nepal's rivers: Indian firms, both private and state-backed, invest heavily in hydropower across the border. India's prime minister, Narendra Modi, is pushing for cross-border energy co-operation in South Asia, in particular to promote the export of Nepali hydropower. Such collaboration could easily become the basis for wider expansion of trade and investment, with water as just one example of growing interactions.

Those who most fear that conflict will follow a lack of fresh water and competition for use of cross-border rivers might note the existence of one study of

263 trans-boundary rivers around the world, by researchers at Oregon State University. It suggested co-operation over rivers is more common than conflict (Rivers and conflict 2008).

Where conflict appears most likely to arise in relation to water in South Asia—as Indian diplomats have suggested—is in fact when the subject is used as a pretext. In Arunachal Pradesh, for example, the root cause of the dispute between India and China is the failure to settle the border and agree on sovereignty over the territory. Water is a sub-set of a much bigger dispute. Jihadi and militant groups in Pakistan, hostile to the existence of India, raise claims that India extracts more water from the Indus river tributaries than is permitted by the two countries' water treaty. This is (so far) not true, but the claim is convenient because it makes the upstream power appear aggressive, and appears, to sympathizers of the groups, to justify subsequent threats to bomb dams in Indian territory. Yet if the pretext were not water, some other excuse could be raised instead. Much the same can be said of how the issue of water is used by officials of the two countries, as Chellaney himself concedes in passing, "In reality, the India-Pakistan water wrangling is just regional politics by another name".

In many cases politicians take advantage of the fact that many South Asians hold a bitterly gloomy view of their neighbours' behaviour on transboundary water issues (and perhaps encourage them to be more hostile still). According to an opinion survey of almost 500 individuals considered experts in water, conducted on behalf of Chatham House, a London think-tank, in 2013, 78 % of Pakistani respondents, 82 % of Bangladeshis and 76 % of Nepalese had a negative view of Indian behaviour over water. Indians, generally, had more mixed views of their neighbours (see www.chathamhouse.org/publication/attitudes-water-south-asia).

5 Conclusion

Many commentators and actors in South Asia are right to warn that water scarcity is a great and growing problem for the region, one already troubled by discord. As competition and then conflict could follow from the scarcity of any finite, valuable, resource, South Asia could be vulnerable to new conflict that arises given a lack of freshwater. However most solutions to the growing problem of scarcity—changing practices by farmers and irrigation, better co-operation by sub-national actors, investment in irrigation and pipes, improved technology and the proper pricing of water—in fact arise as matters of domestic policy, not internationally. The scarcity of water in Pakistan, for example, is not a result of Indian extraction of excess quantities of water upstream of the Indus River, but results from excessive Pakistani extraction and from leakage. In addition, it is not sufficient to assume that the scarcity of a given resource in itself will trigger conflict rather than, more hopefully, become the basis for co-operation. Existing signs of co-operation in South Asia, such as long-lasting water treaties, emerging energy trading, or Indian investment in hydropower in Nepal, suggest prospects for talks on cross-border rivers being a source of more stability, not less. Actors that in any case seek conflict may use the subject of water as a pretext. That aside, scarcity of water could as easily be the means to reduce discord in South Asia as increase it.

Acknowledgments Thanks to Dr. Anne Hammerstad and Indrani Bhattacharya for comments and suggestions on this.

References

Chellaney B (2011) Water: Asia's new battleground. HarperCollins, New Delhi
India's climate: monsoon, or later (2012) The Economist, July 29 2012
Rivers and conflict: streams of blood, or streams of peace (2008) The Economist, May 1 2008
South Asia's water: unquenchable thirst (2011) The Economist. Interview with the author, Nov 19 2011

Index

Note: Page numbers followed by *f* and *t* indicate figures and tables, respectively

A.K. Biswas and C. Tortajada (eds.), *Water Security, Climate Change and Sustainable Development*, Water Resources Development and Management, DOI 10.1007/978-981-287-976-9

FSC
www.fsc.org
MIX
Papier | Fördert gute Waldnutzung
FSC® C083411

Kolibri 360 GmbH
Ferdinand-Jühlke-Straße 7
99095 Erfurt, Deutschland
produktsicherheit@kolibri360.de